The Unique Properties of 4D Spiral Spacetime
Toryx

First Edition

by
Vladimir B. Ginzburg

Cover by Eugene B. Ginzburg

Helicola Press
Division of IRMC, Inc.
612 Driftwood Drive
Pittsburgh, Pennsylvania, 15238

The Unique Properties of
4D Spiral Spacetime
Toryx

First Edition

Published by
Helicola Press
Division of IRMC, Inc.
612 Driftwood Drive
Pittsburgh, Pennsylvania
USA

Printed in the United States of America

ISBN: 978-0-9671432-8-6 (Perfect bound)

Current printing (last digit)
10 9 8 7 6 5 4 3 2 1

Contents

My special thanks go to the members of my family: to my still young, gorgeous, friendly and energetic grandson Alex, to my beautiful and talented grandson Asher who proofread Introduction of this book; to my daughter Ellen with whom I brainstormed some ideas related to my theory and who proofread Summary of this book; to my son Gene whose comments about a multi-level universe made in December of 1992 had triggered my interest in the development of the theory described in this book, and who also designed covers of all my books and helped me to produce them; to my brother Paul who patiently followed my research in this field and provided me with his valuable comments and corrections, and, finally, to my wife Tanya, whose advice and assistance in editing my books were invaluable. Many great memories about my wonderful late parents and very wise grandparents provided me with a needed moral support in writing my books.

ACKNOWLEDGEMENTS

I am grateful to several scientists for making their valuable and stimulating comments on my idea of multi-level Universe and also on my spiral spacetime model. Among them are:

Professor Carlo Rovelli (Department of Physics and Astronomy of the University of Pittsburgh),
Professor Gregory M. Townsend (Department of Physics of the University of Akron),
Professor Clifford Taubes (Department of Mathematics, Harvard University)
Professor Rudolph Hwa (Department of Physics, University of Oregon),
Dr. Blair M. Smith (Innovative Nuclear Space Power and Propulsion Institute, University of Florida),
Professor Edward F. Redish (Department of Physics, University of Maryland),
Professor David F. Measday (Department of Physics and Astronomy, University of British Columbia, Canada),
Dr. Eric Carlson (Department of Physics & Astronomy, Wake Forest University),
Professor Warren Siegel (Department of Physics and Astronomy, Stony Brook University).
Dr. Charles E. Hyde-Wright, Professor of Physics, Old Dominion University,
Dr. John G. Fetkovich, Professor Emeritus (Department of Physics, Carnegie Mellon University),
Dr. Yan Corlett, York University, Canada,
Dr. Igor Sokolov, Assistant Professor of Department of Physics, Clarkson University) and
Dr. Thomas Love, California State University at Dominguez Hills.

Thanks to the recommendations made by Dr. Akhlesh Lakhtakia (Department of Engineering and Mechanics, the Pennsylvania State University), I was able to publish the first three papers describing the earliest versions of my theory in *Speculations in Science and Technology* in 1996-1998. I appreciate very much my interesting discussions with late Marvin Solit, Director of Foundations for New Directions, and his ideas about possible space orientations of toryces. I will always remember very stimulating discussions with the late Distinguished Professor Eli Gorelik of the University of Pittsburgh, Pennsylvania about a crystal structure of a nucleon core.

I value greatly my two meetings with Dr. Arlie Oswald Petters, the Benjamin Powell Professor and Professor of Mathematics, Physics and Business Administration at Duke University held at the Duke University in October 2015. After an open-minded review of a basic concept of my theory, Dr. Petters offered me several practical recommendations on how to introduce it to the academic world. Finally, let me express my gratitude to Dr. John Kern II, Chair, Dr. Anna Haensch and Ms. Larisa Shtrahman, Instructor of Mathematics (all from Department of Mathematics & Computer Science of Duquesne University, Pittsburgh, PA), for providing me with an opportunity to conduct a seminar on mathematical aspects of my theory in their department on February 15, 2017. I also appreciate an invitation to present my theory at the upcoming Materials Science & Technology (MS&T) 2017 Conference to be held in Pittsburgh, PA.

Vladimir B. Ginzburg

This book is dedicated to the German-Swiss physicist Albert Einstein who explained gravity as a result of interactions of celestial bodies with spacetime described by a set of his ten field equations. He also opened a door to a search for a more inclusive spacetime helping us to understand many other phenomena of both micro- and macro-worlds.

This book is also dedicated to Walter Russell, the American self-educated scientist, architect and artist who envisioned the spacetime origin of the Universe in his book:

> *That the man calls matter, or substance, has no existence whatsoever. So-called matter is but waves of the motion of light, electrically divided into opposed pairs, then electrically conditioned and patterned into what we call various substances of matter. Briefly put, matter is the motion of light, and motion is not substance. It only appears to be. Take motion away and there would not be even the appearance of substance.*

Walter Russell

The author of *The Secret of Light.*

INTRODUCTION

According to the title of this book, the toryx is a four-dimensional (4D) spiral spacetime. It means its properties are described by three space plus one time parameters. Part 1 of this book presents properties of toryces in abstract mathematical terms. Part 2 shows several applications of toryces for mathematical modeling of properties of entities of both micro- and macro-worlds. This book further confirms a main proposition of the author's Universal Space Theory (UST) that the toryx has all attributes required to be a prime element of nature.

I1. Discovering the Toryx

There are many ways to discover the toryx. One of the ways includes two steps: first, a modification of the de Broglie's model of an electron inside an atom accompanied by a standing wave (see left side of Figure I1.1) and, second, the introduction of *three toryx spacetime postulates*.

1. Modification of the de Broglie's model - As shown at the right side of Figure I1.1, in the toryx, the orbital path of electron becomes its circular *leading string* with the radius r_1 and propagating with the velocity of electron V_1. The standing wave accompanying the electron of the de Broglie's model is replaced with a toroidal *trailing string* with the radius r_2 propagating synchronously with the circular leading string. So, the translational velocity of trailing string V_{2t} is equal to the velocity of electron V_1.

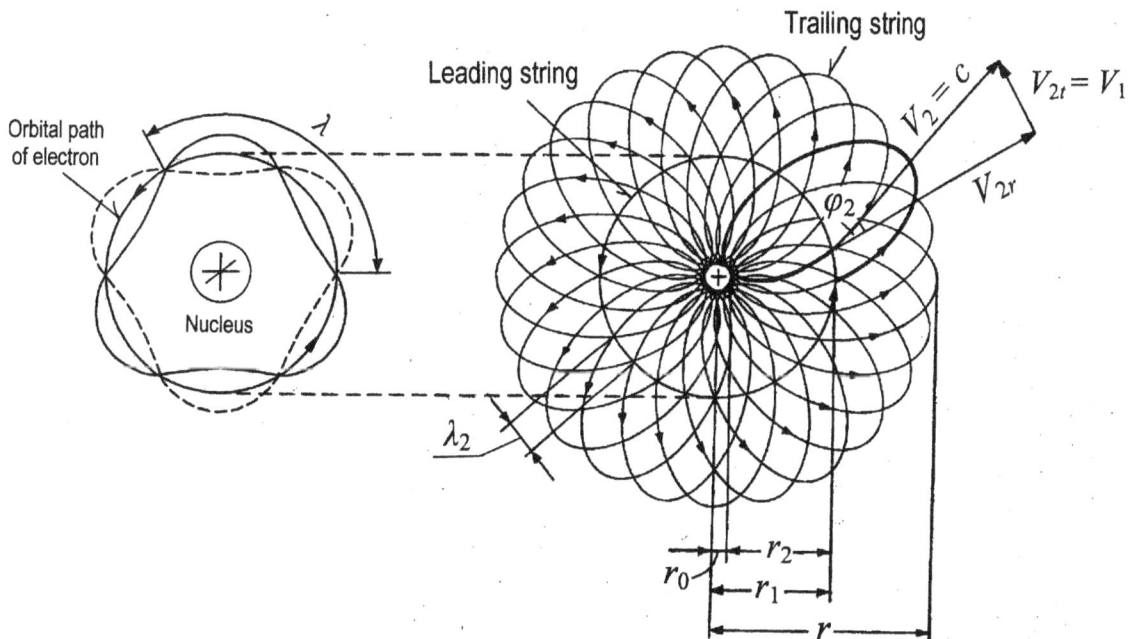

Figure I1.1. Conversion of de Broglie's model of electron (left) into a toryx (right).

2. Introduction of three toryx spacetime postulates – These postulates limit the toryx degrees of freedom as shown in Exhibit I1, while allowing for the radius r_1 of its leading string to decrease from positive to negative infinity.

Exhibit I1. Toryx spacetime postulates in absolute units.

- The length of one winding of toryx trailing string L_2 is equal to the length of one winding of toryx leading string L_1:

$$L_2 = L_1 = 2\pi r_1 \qquad (I1\text{-}1)$$

- The toryx eye radius r_0 is constant:

$$r_0 = r_1 - r_2 = const. \qquad (I12)$$

- The spiral velocity of toryx trailing string V_2 is constant at each point of its spiral path:

$$V_2 = \sqrt{V_{2t}^2 + V_{2r}^2} = c = const. \qquad (I1\text{-}3)$$

The above shown limits are needed to derive equations establishing the relationships between the toryx spacetime parameters, including the wavelength λ_2, the number of windings w_2 of its trailing string and also the velocities, frequencies and periods of its leading and trailing strings. The selected toryx spacetime postulates yielded the simplest way of deriving the toryx spacetime equations.

I2. Analysis of Toryx Spacetime Equations

Analysis of derived toryx spacetime equations reveals its unique topological properties. Figure I2.1 shows that as the radius of the toryx leading string decreases from positive to negative infinity, the steepness angle of trailing string φ_2 increases from 0^0 to 360^0.

Consequently, four kinds of topologically-inverted toryces are formed:

- **Real negative toryces** in which *radius of trailing string* becomes inverted at $\varphi_2 = 90^0$
- **Real positive toryces** in which *radius of spherical boundary* becomes inverted at $\varphi_2 = 180^0$
- **Imaginary positive toryces** in which *radius of leading string* becomes inverted at $\varphi_2 = 270^0$
- **Imaginary negative toryces** with *entire toryx* becoming inverted at $\varphi_2 = 360^0$.

$$V \to \pm 0$$

Real inversion toryx

$b_1 = 1$

$b_1 = 2$

$b_1 = \frac{2}{3}$

Real positive toryces

Real negative toryces

$120°$ $90°$ $60°$

$150°$

Real toryces

$30°$

$b_1 \to -\infty$ $b_1 \to +\infty$

$180°$ φ_2 $0/360°$

Positive inversion toryx

$V = +1$

Negative inversion toryx

$V = -1$

Imaginary toryces

$210°$ $330°$

$b_1 = \frac{1}{2}$

$240°$ $270°$ $300°$

Imaginary positive toryces

Imaginary negative toryces

$b_1 = \frac{1}{3}$

$b_1 = \to +0$ $b_1 = \to -0$

$b_1 = -1$

Imaginary inversion toryx

$$V \to \pm \infty$$

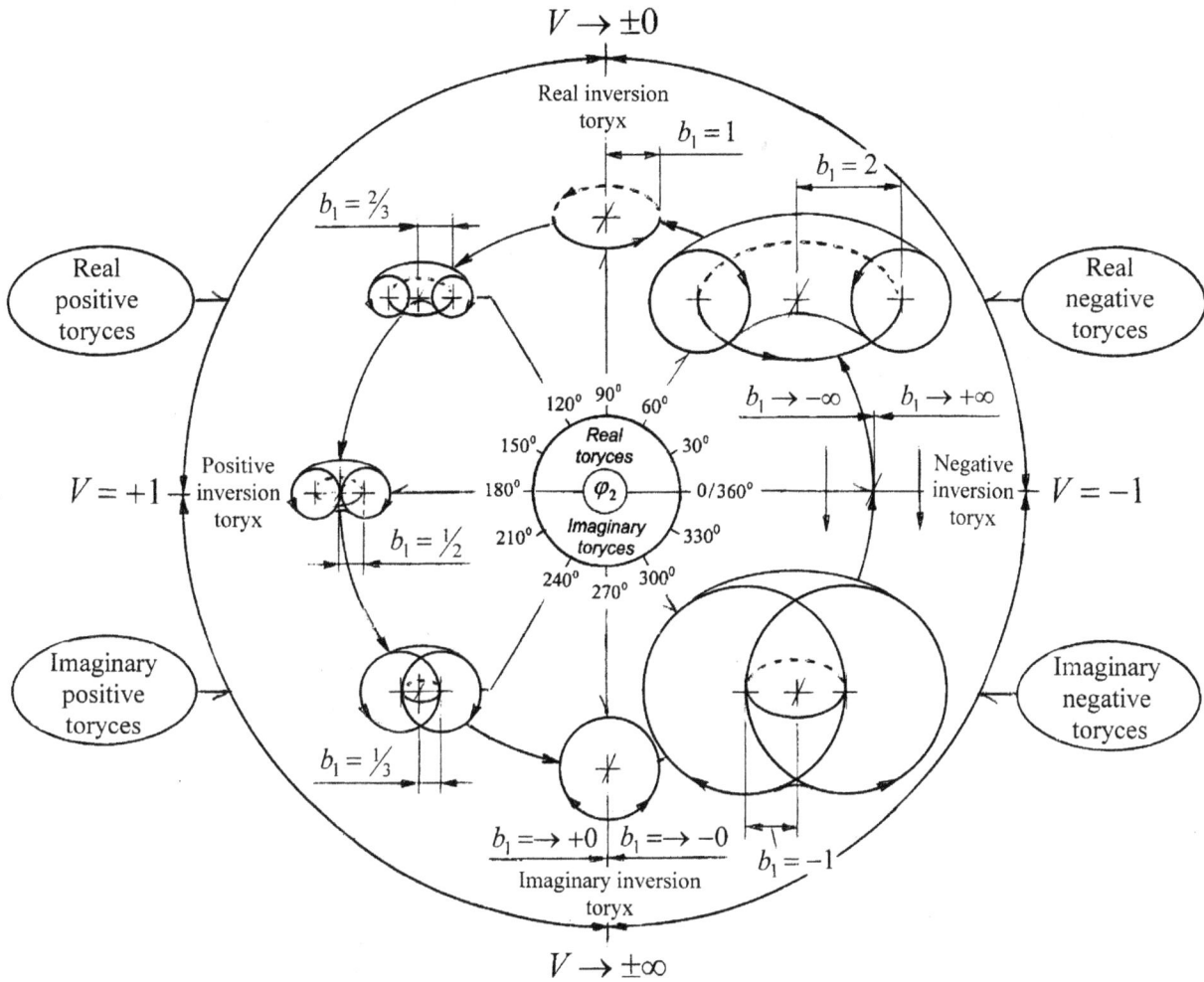

Figure I2.1 Metamorphoses of toryx leading and trailing strings as a function of the steepness angle of trailing string φ_2.

Toryces are divided into four main groups according to their reality R and vorticity V as shown in Table I2.

Table I2. Main classification of toryces.

Toryx name	φ_2	$R = \pm\sqrt{\dfrac{r}{r_0}}$	$V = -\dfrac{r_2}{r_1}$
Real negative	$0° - 90°$	Real	$(-)$
Real positive	$90° - 180°$	Real	$(+)$
Imaginary positive	$180° - 270°$	Imaginary	$(+)$
Imaginary negative	$270° - 360°$	imaginary	$(-)$

I3. Quantum States of Toryces

Toryces change their dimensions in quantum steps by excitation and oscillation. During excitation of a toryx the radius of its leading string r_1 increases, while its eye radius r_0 remains constant. During oscillation of a toryx, its radius of leading string r_1 and its eye radius r_0 change proportionally as shown in Fig. I3.1.

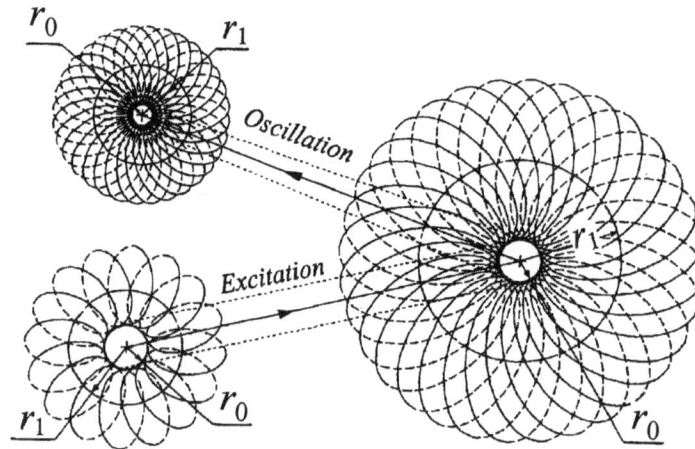

Figure I3.1. Excitation and oscillation of a toryx.

The derived quantization equations of excited toryces are based on two proposed limitations of degrees of freedom of real negative *lambda, harmonic* and *golden toryces* shown in Exhibit I3.1.

Exhibit I3.1. Limitations of degrees of freedom of excited toryces.

The relative radius of leading string of real negative toryx b_1 is equal to:		
Lambda toryx	**Harmonic toryx**	**Golden toryx**
$b_1 = z = 2(n\Lambda)^m$	$b_1 = z = 2 + n$	$b_1 = z = 2 + n/\phi$

In Exhibit I3.1:

        z = toryx quantization parameter
        $m \rightarrow 0, 1, 2, ..$, toryx exponential excitation quantum states
        $n \rightarrow 0, 1, 2, ..$, toryx linear excitation quantum states
        $\Lambda = 137, ...$, *toryx quantization constant*.

The oscillation of a toryx is a function of the *toryx oscillation factor* Q_q calculated based on the proposed limitation of its degree of freedom shown in Exhibit I3.2.

Exhibit I3.2. Limitations of degrees of freedom of oscillated toryces.

The toryx oscillation factor Q_q is equal to:

$$Q_0 = 1; \quad Q_q = 3\left(\frac{\Lambda}{2(q-1)}\right)^{q-1}$$

where
$$q = 0, 1, 2, \ldots, \textit{\textbf{toryx oscillation quantum states}}.$$

The values of Q_q are shown in Figure I3.2 and Table I3.

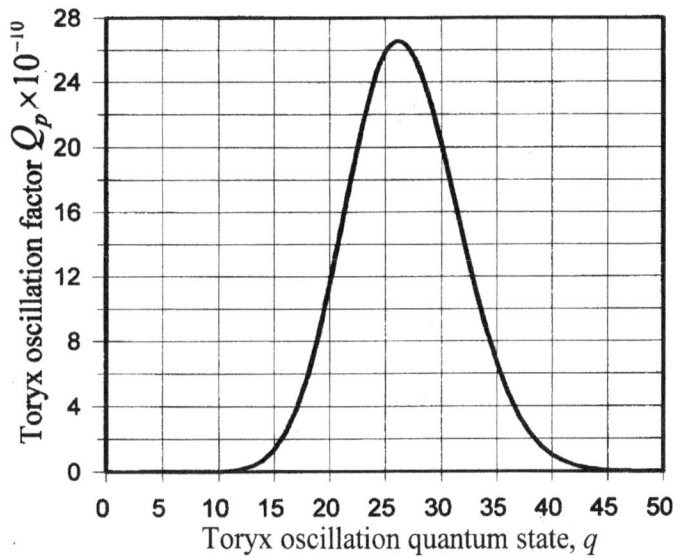

Figure I3.2. Toryx oscillation factor Q_q as a function of the toryx oscillation state q.

Table I3. Toryx oscillation factor Q_q as a function of the toryx oscillation quantum state q.

q	0	1	2	3	4	5	6
Q_q	1.000	3.000	205.500	3511.1875	35713.236	258014.18	1447851.734

The toryces exist in quantum states. When a toryx quantum state changes from a higher to a lower level, it emits a helical spacetime called the ***helyx***.

The UST shows that the above described toryx topology provides them with a sufficient variety of spacetime properties that are necessary to model various physical properties of entities of both micro- and macro-worlds.

I4. Formation of Elementary Matter Particles

According to the UST, the limits of the toryx spacetime are defined by the Planck length l_p below which the toryx spacetime properties are no longer governed by the toryx spacetime postulates, but by the rules of uncertainty of **quantum vacuum**. Consequently, the primordial toryces are produced spontaneously in quantum vacuum in accordance with the proposed **toryx uncertainty principle**. We call the spontaneously created toryces the **virtual toryces**.

Figure I4.1 shows ranges of the toryx relative radii of spherical boundaries b within which the spacetime ceases to exist and replaced by virtual toryces of quantum vacuum. The relative spacetime length b_s corresponding to the limits of spacetime is equal to $l_p/2\pi$.

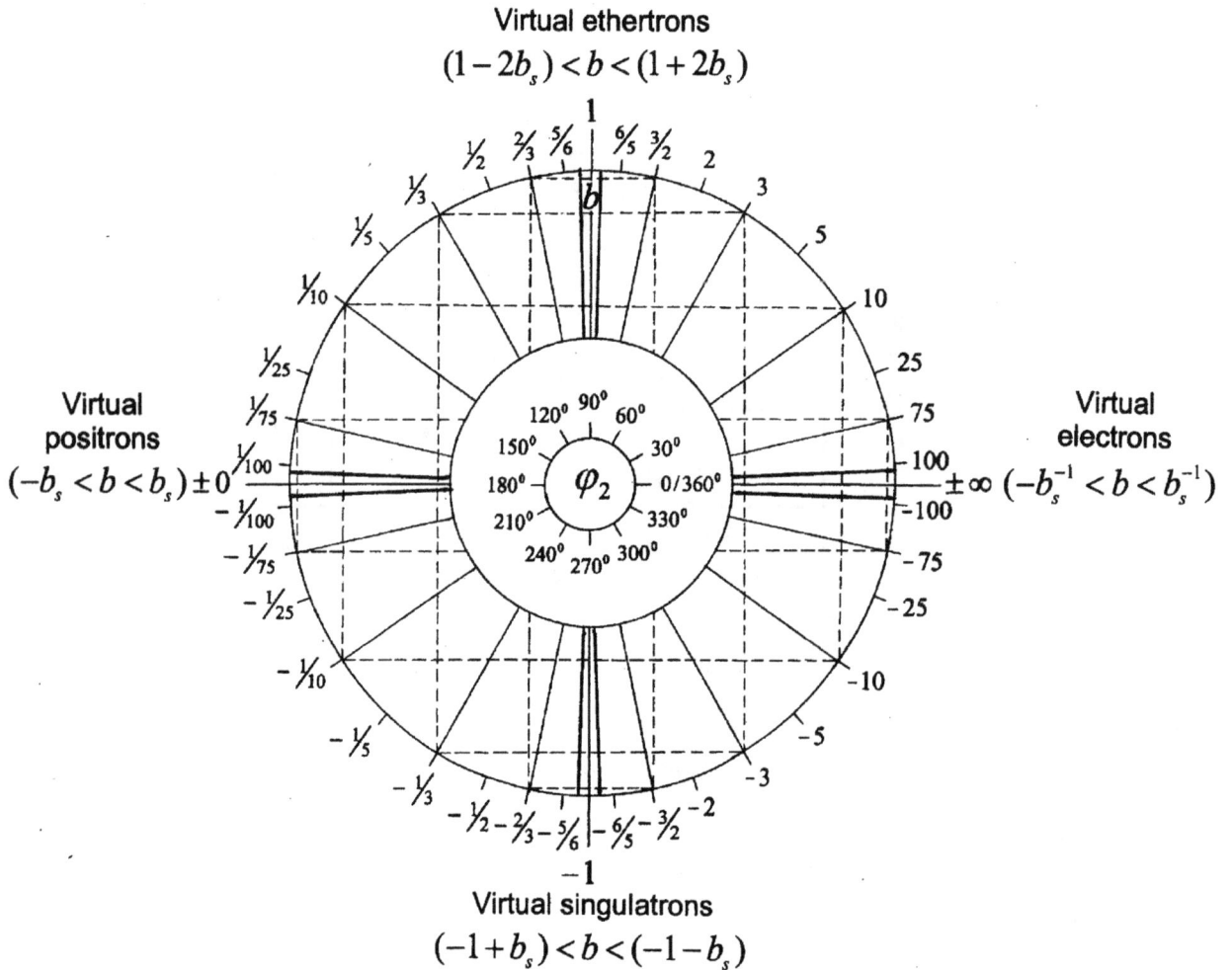

Figure I4.1. Ranges of the toryx relative radii of spherical boundaries b within which the spacetime ceases to exist and replaced by the virtual particles of quantum vacuum.

I5 From Quantum Vacuum to Elementary Matter Particles

The virtual toryces obeying to three toryx spacetime postulates can become the prime elements of nature called *micro-toryces*, or simply toryces. Four basic elementary particles (trons) are formed from quantum vacuum by the unification of adjacent toryces: *electrons, positrons, ethertrons* and *singulatrons* as shown in Figs. I5.

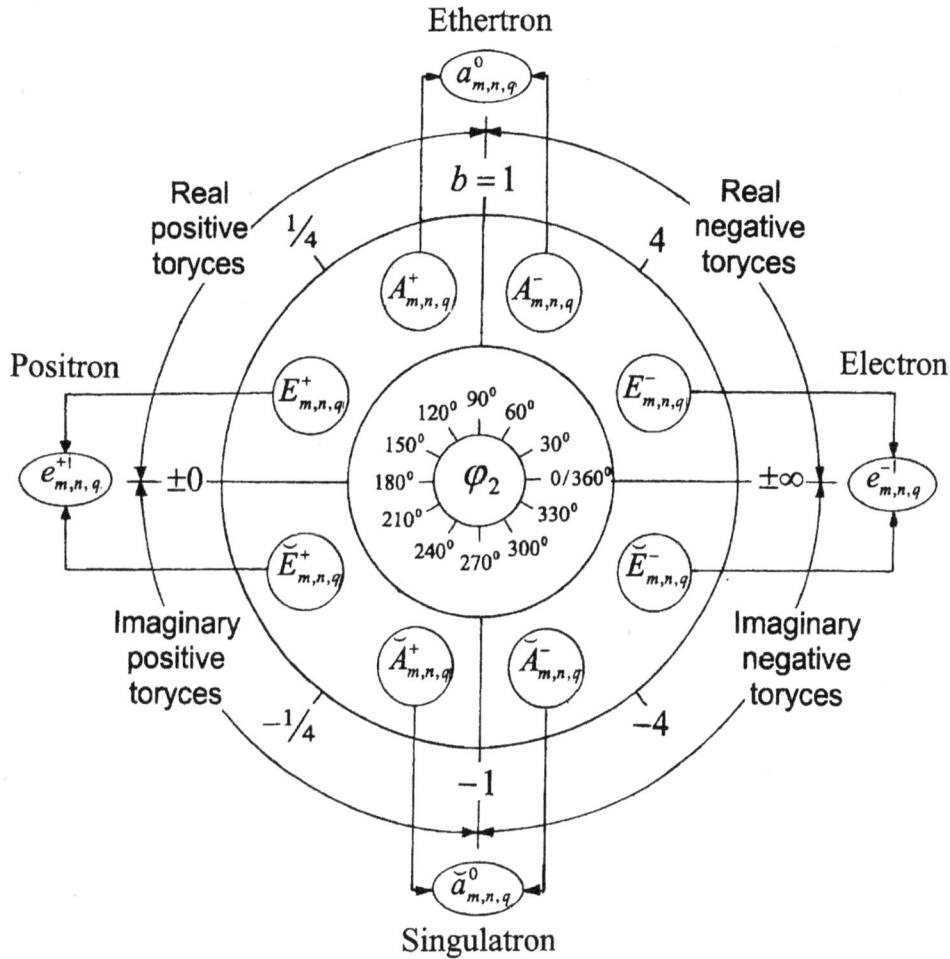

Figure I5. Formation of mutually-polarized excited lambda trons.

As shown in Fig. I5.1:

- Electrons are made up of real and imaginary negative toryces
- Positrons are made up real and imaginary positive toryces
- Ethertrons are made up of real negative and positive toryces
- Singulatrons are made up of imaginary negative and positive toryces.

Imaginary toryces are responsible for absorption of spacetime, while real toryces are responsible for release of spacetime. The stability of reality-polarized trons is governed by the proposed ***tron polarization conservation law.***

I6. Generic Spacetime Properties of Toryces

Spacetime properties – Toryces are the smallest entities of our Universe which properties are described in ***objective spacetime terms*** comprehendible by all civilized inhabitants of the Universe. The spacetime properties can be found in all entities of the Universe. Take the spacetime properties away and the Universe, as we know it, will cease to exist.

Self-preservation – Toryces are the smallest stable entities of our Universe that are able to self-preserve their existence. All stable entities of the Universe exist because they are able to self-preserve themselves. Take the self-preservation away and the Universe, as we know it, will cease to exist.

Motion – Toryces are the smallest entities of our Universe whose leading and trailing strings are in a state of continuous motion. All entities in the Universe are in the motion state. Take the motion away and the Universe, as we know it, will cease to exist.

Spirality – Toryces are the smallest entities of our Universe whose trailing strings propagate along a spiral path. Spirality pervades the entire Universe, including the paths of all celestial bodies. It is also at the core of a structure of DNA. Take spirality away and the Universe, as we know it, will cease to exist.

Propagation with velocity of light – Toryces are the smallest entities of our Universe whose trailing strings propagate along their spiral paths with velocity of light. Take the velocity of light away and the Universe, as we know it, will cease to exist.

Limitation of spacetime freedom – Toryces are the smallest entities of our Universe with limited spacetime freedom. The spacetime freedom is limited in all entities of the Universe. Take the limitation of spacetime freedom away and the Universe, as we know it, will cease to exist.

Planetary motion – Toryces are the smallest entities of our Universe in which their leading strings follow the spiral spacetime law of planetary motion. Atomic electrons and planets follow a law of planetary motion. Take the law of planetary motion away and the Universe, as we know it, will cease to exist.

Polarization – Toryces are the smallest entities of our Universe that exist in polarized states. Atoms are made up of electrically-polarized electrons and nuclei, while many complex entities, including DNA, are made up of chemically-polarized acids and bases. Take the polarization away and the Universe, as we know it, will cease to exist.

Expansion & contraction – Toryces are the smallest entities of our Universe able to expand and contract. This capability can be found in all entities of the Universe. Take the expansion and contraction away and the Universe, as we know it, will cease to exist.

Quantization – Toryces are the smallest entities of our Universe that exist in spacetime excitation and oscillation quantum states. This property extends to all elementary particles making up atoms. Take the quantization away and the Universe, as we know it, will cease to exist.

Stable coexistence – Toryces are the smallest entities of our Universe that can stably coexist with their respective oppositely-polarized toryces. Electrons and protons stably coexist in atoms, while acids and bases stably coexist in DNA. Take stable coexistence away and the Universe, as we know it, will cease to exist.

Absorption & release of spacetime – Toryces are the smallest entities of our Universe able to absorb and release spacetime to sustain their existence. All entities of the Universe sustain their existence by absorption and release of spacetime, or energy, in physical terms. Take this capability away and the Universe, as we know it, will cease to exist.

I7. Macro-Toryces

Each body of the macro-world can be thought as an assembly of elementary matter particles, atoms, chemical compositions and other more complex components with all of them made up of polarized micro-toryces. These assemblies of polarized toryces form the ***macro-toryces*** intimately associated with each body. Figure I7.1 shows a macro-toryx associated with the central body A and encompassing the satellite body B.

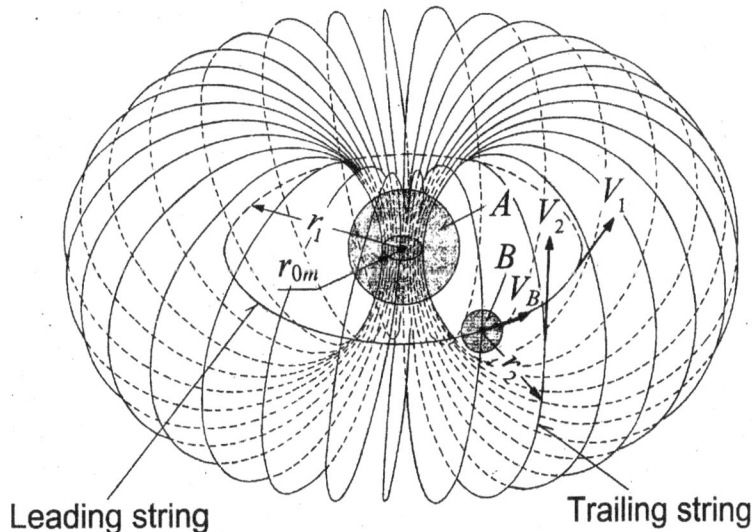

Leading string   Trailing string

Figure I7.1. Central body A with a macro-toryx encompassing the satellite body B.

Similarly to the micro-toryx (Fig. I7.1), the macro-toryx contains two strings, a ***leading string*** and a ***trailing string***. Leading string is circular; it is moving at the velocity V_1 along a circle with the radius r_1. Trailing string is toroidal with the radius r_2. The toroidal trailing string propagates around the circular leading string with the spiral velocity V_2. Notably, micro-toryces and macro-toryces have the same structures and their spacetime properties are described by the same equations, except for the radius of the toryx eye r_0.

The UST proposes that the macro-trons associated with celestial bodies are responsible for the interactions between these bodies. Figure I7.2 shows two adjacent celestial bodies A and B and the respective macro-toryces A and B associated with these bodies. Let the macro-toryx A to be associated with the body A having the mass m_A. The macro-toryx circular leading string with the radius r_{1A} propagates around a center of the body A with the orbital velocity V_{1A}, while its toroidal trailing string with the radius r_{2B} propagates synchronously with the leading string. Located at the distance r_{1A} from the center of the body A is a center of the body B having the mass m_B. The body B moves around the center of the body A along a circular path with the radius $r_{1B} = r_{1A}$ with the orbital velocity V_B.

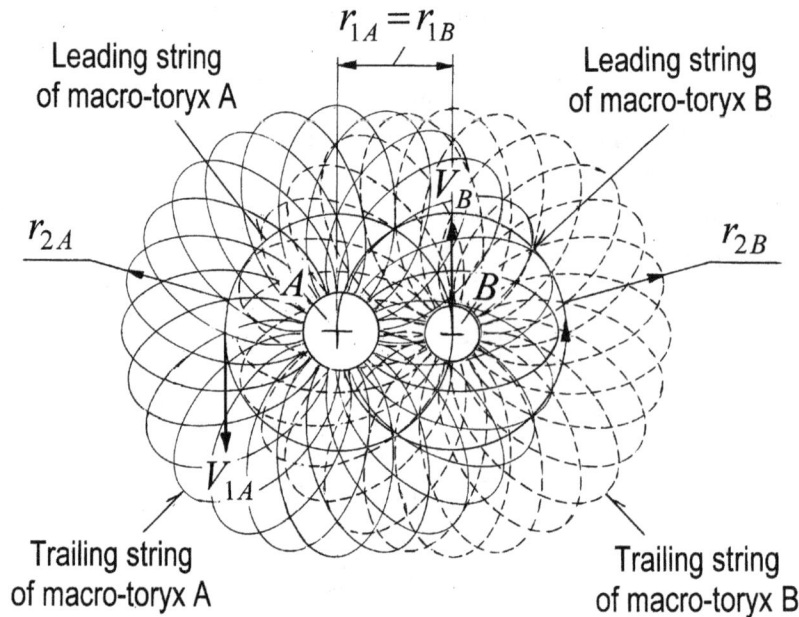

Figure I7.2. Two adjacent celestial bodies A and B with macro-toryces A and B associated with these bodies.

The behavior of the body B depends on a relationship between its orbital velocity V_B and the spiral velocity of leading string V_{1A} of the macro-toryx A of the body A. When these velocities are not exactly the same, it means the body B does not follow the *spacetime law of planetary motion*. Consequently, the body B will move in a radial direction either towards to or away from the body A with the *spacetime acceleration* a_{bs} that is directly-proportional to the difference between squares of the orbital velocity V_B of the body B and the spiral velocity of macro-toryx leading string V_{1A} associated with the body A and inversely-proportional to the distance between the bodies A and B.

Consequently, it is possible to discover two general spacetime laws applicable to a wide range of sizes and densities of celestial bodies, including neutron stars, pulsars and black holes, and also applicable to a wide range of distances between celestial bodies:

- The *spacetime law of gravitation* for which the Newton's universal law of gravitation is a particular case.

- The *spacetime law of planetary motion* for which the Kepler's third law of planetary motion is a particular case.

I8. The Spacetime Origin of the Universe

Physical properties of toryces are directly related to their spacetime properties.

What current theories of physics describe as matter, field, charge, mass, electromagnetism and gravity are merely metamorphoses of spiral spacetimes.

All spacetime parameters of micro-toryces and elementary matter particles can be calculated based on the following spacetime constants:

Velocity of light $c = 2.997\ 924\ 58 \times 10^{08}\,\text{m/s}$

Micro-toryx eye radius $r_0 = 1.408\ 970\ 164 \times 10^{-15}\,\text{m}$

Toryx quantization constant $\Lambda = 137$.

The rest is commentary, as described in this book.

Notes

PART 1

Abstract
Mathematics
of a Toryx

1. TORYX BASIC STRUCTURE & PARAMETERS

1.1 Toryx Basic Structure

Toryx is a single-level 4d helicola. Its basic structure consists of a double-circular *leading string* and a double-toroidal *trailing string*, both residing inside a *spherical boundary* as shown in Figs. 1.1.1 – 1.1.3. Notice that for the sake of simplicity these figures show only one branch of the double-circular leading and double-toroidal trailing strings.

Toryx leading string – The toryx double-circular leading string appears like two circular traces with the radius r_1 left by the moving points m and n shown in Figure 1.1.1. It can be thought as a dynamic double-helical spiral in which the translational velocity V_{1t} is equal to zero, so the rotational velocity V_{1r} is equal to the spiral velocity V_1. Thus,

$$V_{1t} = 0 \qquad\qquad (1.1\text{-}1)$$

$$V_{1r} = V_1 \qquad\qquad (1.1\text{-}2)$$

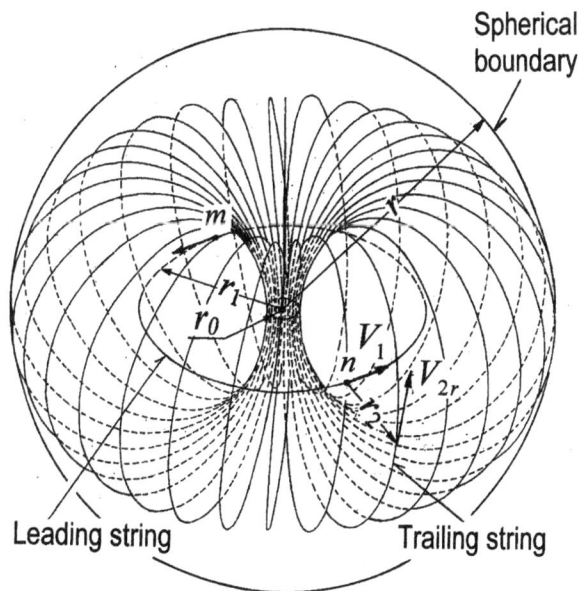

Figure 1.1.1. Isometric view of a toryx.

Toryx trailing string – The toryx double-toroidal trailing string has the radius r_2 and a circular opening at the toryx center with the radius r_0 called the ***toryx eye***. It can be thought as a dynamic double-toroidal spiral in which each branch propagates along its toroidal spiral path with the spiral velocity V_2 that has two components, the translational velocity V_{2t} and the rotational velocity V_{2r}, with all three velocities related to each other by the Pythagorean Theorem:

$$V_2 = \sqrt{V_{2t}^2 + V_{2r}^2} \tag{1.1-3}$$

The trailing string propagates synchronously with the leading string. Therefore, the translational velocity of trailing string V_{2t} is equal to the rotational velocity V_{1r} and the spiral velocity V_1 of leading string as given by:

$$V_{2t} = V_{1r} = V_1 \tag{1.1-4}$$

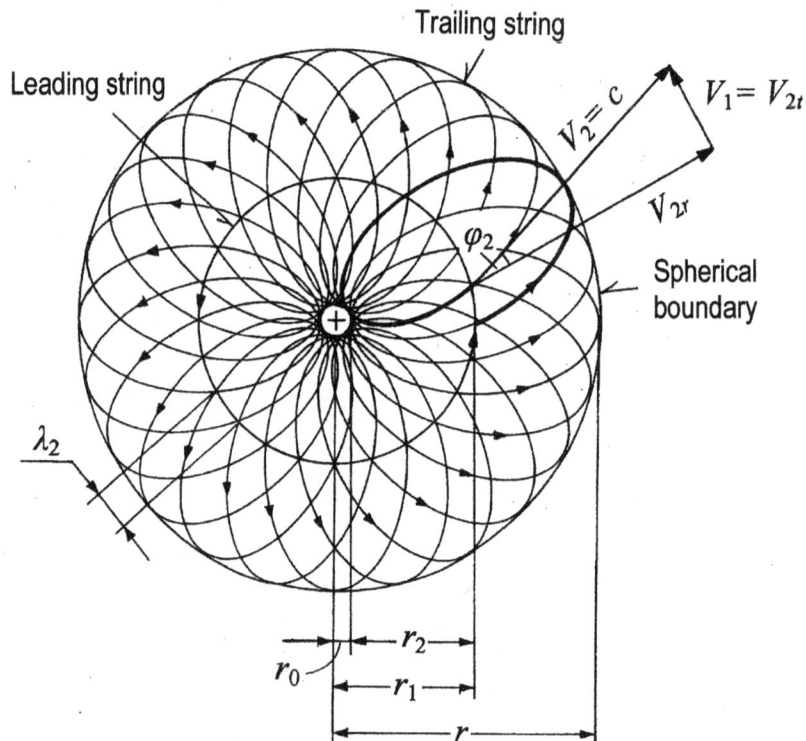

Figure 1.1.2. Top view of a toryx.

In Fig. 1.1.2, φ_2 is the ***steepness angle*** of toryx trailing string corresponding to the middle point a of toryx trailing string.

The radius of spherical boundary r is equal to:

$$r = r_1 + r_2 \qquad (1.1\text{-}5)$$

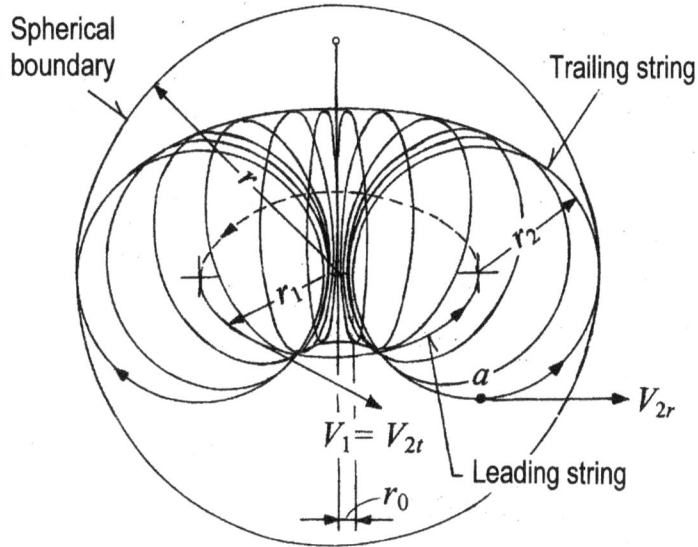

Figure 1.1.3. Cross-section of a toryx.

Toryx spins – Toryx has two spins, leading and trailing, with both of them defined by the right-hand rule as shown in Fig. 1.1.4. The toryx leading and trailing spins depend on the directions of the rotational velocities of toryx leading string V_{1r} and toryx trailing string V_{2r}.

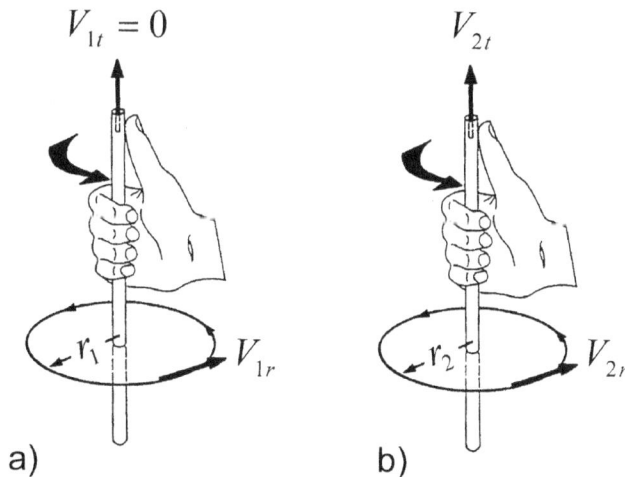

Figure 1.1.4. The toryx leading spin (a) and trailing spin (b) defined by the right-hand rule.

1.2 Toryx Spacetime Parameters in Absolute Units

The values of some toryx spacetime parameters are dependent on the position of a point of measurement of these parameters along the toryx trailing string. In the nomenclature of toryx parameters shown below, the symbols of the position-dependent parameters marked with star (*) correspond to the middle point a of toryx trailing string.

f_0 = toryx base frequency
f_1 = frequency of toryx leading string
f_2 = frequency of toryx trailing string
L_1 = spiral length of one winding of toryx leading string
L_2 = spiral length of one winding of toryx trailing string
L_0 = circular length of toryx eye
r = radius of toryx spherical boundary
r_0 = toryx eye radius
r_1 = radius of toryx leading string
r_2 = radius of toryx trailing string
T_1 = period of toryx leading string
T_2 = period of toryx trailing string
V_1 = spiral velocity of toryx leading string
V_{1t} = translational velocity of toryx leading string
V_{1r} = rotational velocity of toryx leading string
V_2 = spiral velocity of toryx trailing string
V_{2t} = translational velocity of toryx trailing string*
V_{2r} = rotational velocity of toryx trailing string*
w_1 = the number of windings of toryx leading string
w_2 = the number of windings of toryx trailing string
λ_1 = wavelength of toryx leading string
λ_2 = wavelength of toryx trailing string*
φ_1 = steepness angle of toryx leading string
φ_2 = steepness angle of toryx trailing string*.

1.3 Toryx Spacetime Postulates in Absolute Units

The toryx spacetime postulates include three fundamental equations limiting the degrees of freedom of several toryx parameters (see Exhibit 1.3). These postulates provide the simplest way to derive the relationships between all spacetime parameters of toryx. In spite of their outmost simplicity, these postulates provide toryces with amazing spacetime properties, including a capability to exist in four unique topologically-polarized states within the range of the radius of toryx leading strings r_1 extending from negative to positive infinity $(-\infty < r_1 < +\infty)$ as will be described in Chapter 5.

Exhibit 1.3. Toryx spacetime postulates in absolute units.

- The length of one winding of toryx trailing string L_2 is equal to the length of one winding of toryx leading string L_1:

$$L_2 = L_1 = 2\pi r_1 \qquad (1.3\text{-}1)$$

- The toryx eye radius r_0 is constant:

$$r_0 = r_1 - r_2 = const. \qquad (1.3\text{-}2)$$

- The spiral velocity of toryx trailing string V_2 is constant at each point of its spiral path:

$$V_2 = \sqrt{V_{2t}^2 + V_{2r}^2} = c = const. \qquad (1.3\text{-}3)$$

Figure 1.3 shows the toryx spacetime parameters in absolute units corresponding to the middle point a of toryx trailing string (Fig. 1.1.3).

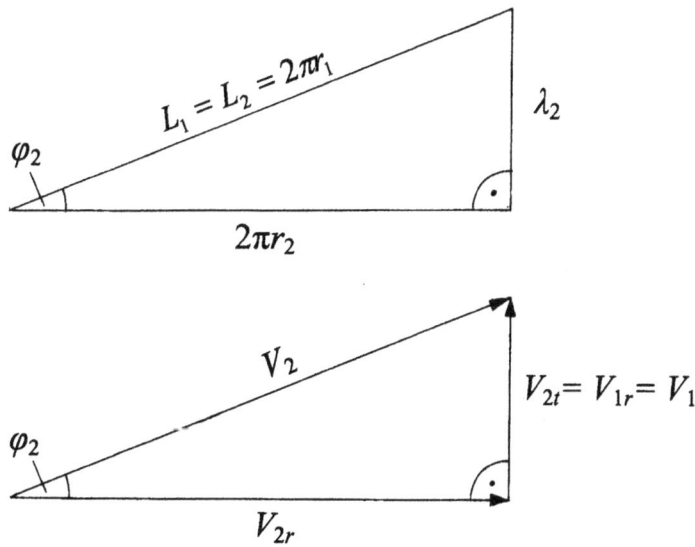

Figure 1.3. Toryx spacetime parameters in absolute units corresponding to the middle point a of toryx trailing string (Fig. 1.1.3).

1.4 Toryx Spacetime Parameters in Relative Units

The toryx spacetime parameters can be simplified by expressing them in relative units in respect to the constant toryx parameters: the toryx eye radius r_0, the velocity of light c and the toryx base frequency f_0 as shown in Table 1.4.

Table 1.4. Toryx relative spacetime parameters.

Toryx relative parameters	Equations
Radius of toryx spherical boundary	$b = r/r_0$ (1.4-1)
Radius of toryx leading string	$b_1 = r_1/r_0$ (1.4-2)
Radius of toryx trailing string	$b_2 = r_2/r_0$ (1.4-3)
Length of toryx leading string	$l_1 = L_1/2\pi r_0$ (1.4-4)
Length of toryx trailing string	$l_2 = L_2/2\pi r_0$ (1.4-5)
Period of one winding of toryx leading string	$t_1 = T_1 f_0$ (1.4-6)
Period of one winding toryx trailing string	$t_2 = T_2 f_0$ (1.4-7)
Spiral velocity of toryx leading string	$\beta_1 = V_1/c$ (1.4-8)
Translational velocity of toryx leading string	$\beta_{1t} = V_{1t}/c$ (1.4-9)
Rotational velocity of toryx leading string	$\beta_{1r} = V_{1r}/c$ (1.4-10)
Spiral velocity of toryx trailing string	$\beta_2 = V_2/c$ (1.4-11)
Translational velocity of toryx trailing string	$\beta_{2t} = V_{2t}/c$ (1.4-12)
Rotational velocity of toryx trailing string	$\beta_{2r} = V_{2r}/c$ (1.4-13)
Frequency of toryx leading string	$\delta_1 = f_1/f_0$ (1.4-14)
Frequency of toryx trailing string	$\delta_2 = f_2/f_0$ (1.4-15)
Wavelength of toryx leading string	$\eta_1 = \lambda_1/2\pi r_0$ (1.4-16)
Wavelength of toryx trailing string	$\eta_2 = \lambda_2/2\pi r_0$ (1.4-17)

The *toryx base frequency* f_0 corresponds to the case when $r_1 = r_0$ and it is equal to:

$$f_0 = \frac{c}{2\pi r_0}$$

(1.4-18)

1.5 Toryx Spacetime Postulates in Relative Units

Exhibit 1.5 shows three toryx spacetime postulates in relative units.

Exhibit 1.5. Toryx spacetime postulates in relative units.

- The relative length of one winding of toryx trailing string l_2 is equal to the relative length of one winding of toryx leading string l_1:

$$l_2 = l_1 \qquad (1.5\text{-}1)$$

- The toryx relative eye radius b_0 is equal to 1:

$$b_0 = b_1 - b_2 = 1 \qquad (1.5\text{-}2)$$

- The relative spiral velocity of toryx trailing string β_2 is equal to 1 at each point of its spiral path:

$$\beta_2 = \sqrt{\beta_{2t}^2 + \beta_{2r}^2} = 1 \qquad (1.5\text{-}3)$$

Figure 1.5 shows the toryx relative spacetime parameters corresponding to the middle point a of toryx trailing string (Fig. 1.1.3).

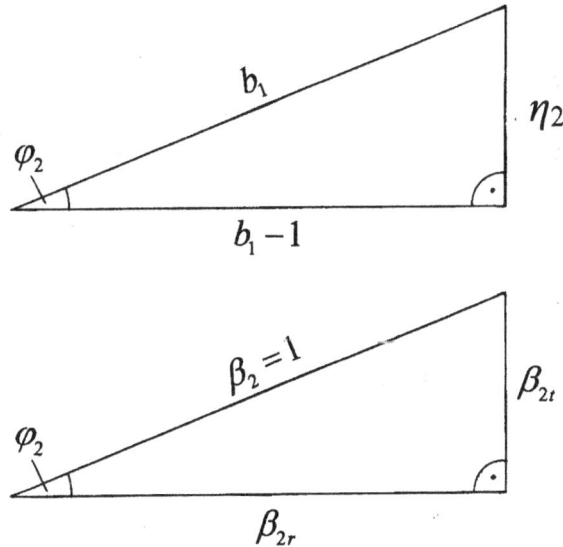

Figure 1.5. Toryx relative spacetime parameters corresponding to the middle point a of toryx trailing string (Fig. 1.1.3).

1.6 Summary of Derived Toryx Equations

Table 1.6.1 provides a summary of derived equations for main relative spacetime parameters of toryx as functions of the relative radius of toryx leading string b_1.

Table 1.6.1. Equations for relative spacetime parameters of toryx leading and trailing strings as functions of the relative radius of toryx leading string b_1.

Relative parameter	Leading string Eq. (a)	Trailing string Eq. (b)
Radius Eq. (1.6-1)	$b_1 = \dfrac{r_1}{r_0}$	$b_2 = \dfrac{r_2}{r_0} = b_1 - 1$
Wavelength Eq. (1.6-2)	$\eta_1 = \dfrac{\lambda_1}{2\pi r_0} = b_1$	$\eta_2 = \dfrac{\lambda_2}{2\pi r_0} = \sqrt{2b_1 - 1}$
Length of one winding Eq. (1.6-3)	$l_1 = \dfrac{L_1}{2\pi r_0} = b_1$	$l_2 = \dfrac{L_2}{2\pi r_0} = b_1$
Steepness angle Eq. (1.6-4)	$\varphi_2 = 0$	$\cos s\varphi_2 = \dfrac{b_1 - 1}{b_1}$
The number of windings Eq. (1.6-5)	$w_1 = 1$	$w_2 = \dfrac{b_1}{\sqrt{2b_1 - 1}}$
Translational velocity Eq. (1.6-6)	$\beta_{1t} = \dfrac{V_{1t}}{c} = 0$	$\beta_{2t} = \dfrac{V_{2t}}{c} = \dfrac{\sqrt{2b_1 - 1}}{b_1}$
Rotational velocity Eq. (1.6-7)	$\beta_{1r} = \dfrac{V_{1r}}{c} = \dfrac{\sqrt{2b_1 - 1}}{b_1}$	$\beta_{2r} = \dfrac{V_{2r}}{c} = \dfrac{b_1 - 1}{b_1}$
Spiral velocity Eq. (1.6-8)	$\beta_1 = \dfrac{V_1}{c} = \dfrac{\sqrt{2b_1 - 1}}{b_1}$	$\beta_2 = \dfrac{V_2}{c} = 1$
Frequency Eq. (1.6-9)	$\delta_1 = \dfrac{f_1}{f_0} = \dfrac{\sqrt{2b_1 - 1}}{b_1^2}$	$\delta_2 = \dfrac{f_2}{f_0} = \dfrac{1}{b_1}$
Period Eq. (1.6-10)	$t_1 = T_1 f_0 = \dfrac{b_1^2}{\sqrt{2b_1 - 1}}$	$t_2 = T_2 f_0 = b_1$

The values of η_2, β_{2t} and β_{2r} correspond to the middle point a of toryx trailing string (Fig. 1.1.3). In Eq. (1.6-4b), $\cos s\varphi_2$ is the toryx trigonometric function. It relates to the trigonometric function $\cos\varphi_2$ of elementary mathematics as follows:

$$\cos s\varphi_2 = \cos\varphi_2 \quad (0 < \varphi_2 < 180^0) \tag{1.6-11}$$

$$\cos s\varphi_2 = 1/\cos\varphi_2 \quad (180^0 < \varphi_2 < 360^0) \tag{1.6-12}$$

The relative radius of toryx spherical boundary b is equal to:

$$b = \frac{r}{r_0} = 2b_1 - 1 \tag{1.6-13}$$

Table 1.6.2 provides a summary of derived equations for relative spacetime parameters of toryx leading and trailing strings as a function of the steepness angle of toryx trailing string φ_2.

Table 1.6.2. Equations for relative spacetime parameters of toryx leading and trailing strings as functions of the steepness angle of toryx trailing string φ_2.

Relative parameter	Leading string Eq. (a)	Trailing string Eq. (b)
Radius Eq. (1.6-14)	$b_1 = \dfrac{1}{1 - \cos s\varphi_2}$	$b_2 = \dfrac{\cos s\varphi_2}{1 - \cos s\varphi_2}$
Wavelength Eq. (1.6-15)	$\eta_1 = 0$	$\eta_2 = \sqrt{\dfrac{1 + \cos s\varphi_2}{1 - \cos s\varphi_2}}$
Length of one winding (Eq. 1.6-16)	$l_1 = \dfrac{1}{1 - \cos s\varphi_2}$	$l_2 = \dfrac{1}{1 - \cos s\varphi_2}$
The number of windings Eq. (1.6-17)	$w_1 = 1$	$w_2 = \dfrac{1}{\sin s\varphi_2}$
Translational velocity Eq. (1.6-18)	$\beta_{1t} = \dfrac{V_{1t}}{c} = 0$	$\beta_{2t} = \sin s\varphi_2$
Rotational velocity Eq. (1.6-19)	$\beta_{1r} = \sin s\varphi_2$	$\beta_{2r} = \cos s\varphi_2$
Spiral velocity Eq. (1.6-20)	$\beta_1 = \sin s\varphi_2$	$\beta_2 = 1$
Frequency Eq. (1.6-21)	$\delta_1 = \sin s\varphi_2(1 - \cos s\varphi_2)$	$\delta_2 = 1 - \cos s\varphi_2$
Period Eq. (1.6-22)	$t_1 = \dfrac{1}{\sin s\varphi_2(1 - \cos s\varphi_2)}$	$t_2 = \dfrac{1}{1 - \cos s\varphi_2}$

The relative radius of toryx spherical boundary b is equal to:

$$b = \frac{1 + \cos s\varphi_2}{1 - \cos s\varphi_2}$$

(1.6-23)

1.7 Toryx Law of Planetary Motion

The first strong indication that the derived toryx spacetime equations were relevant to a real world comes from Eq. (1.6-8a). This equation establishes a relationship between the relative velocity of the toryx leading string β_1 and the relative radius of this string b_1 that is rewritten below:

$$\beta_1 = \frac{V_1}{c} = \frac{\sqrt{2b_1 - 1}}{b_1}$$

(1.7-1)

For the case when $b_1 \gg 1$, the above equation reduces to the form:

$$\beta_1 = \frac{V_1}{c} = \sqrt{\frac{2}{b_1}}$$

(1.7-2)

Let us show that Eq. (1.7-2) expresses the Kepler's third law of planetary motion. The velocity of the toryx leading string V_1 is equal to:

$$V_1 = \frac{2\pi r_1}{T_1}$$

(1.7-3)

From Eqs. (1.6-1a), (1.7-2) and (1.7-3):

$$\frac{2\pi r_1}{T_1 c} = \sqrt{\frac{2r_0}{r_1}}$$

(1.7-4)

we obtain:

$$r_1^3 = \frac{r_0 c^2}{2\pi^2} T_1^2$$

(1.7-5)

Let k to be equal to:

$$k = \frac{r_0 c^2}{2\pi^2}$$

(1.7-6)

From Eqs. (1.3-2) and (1.3-3), both r_0 and c are constant. Therefore, k is constant too. Consequently, Eqs. (1.7-5) and (1.7-6) yield the Kepler's third law of planetary motion:

$$r_1^3 = kT_1^2 \tag{1.7-7}$$

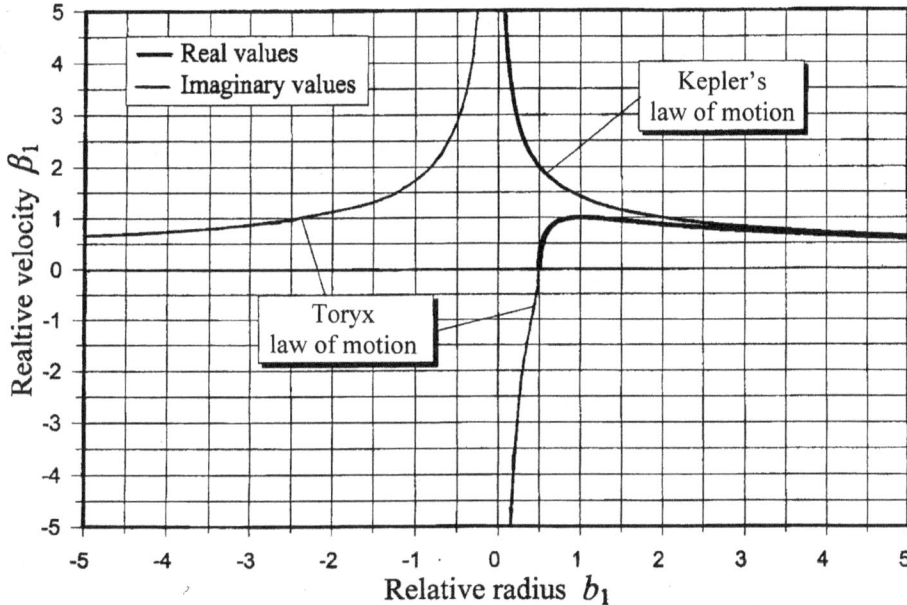

Figure 1.7. Toryx law of motion versus Kepler's third law of planetary motion.

Both Eqs. (1.7-7) and (1.7-2) express the Kepler's third law of planetary motion with only one difference: In Eq. (1.7-7) parameters are expressed in absolute units, while in Eq. (1.7-2) in relative units. Thus, the Kepler's third law of planetary motion expressed by Eq. (1.7-2) is applied for large relative orbital radii ($b_1 \gg 1$) and can be treated as a particular case of a what we called the *Toryx law of planetary motion* that is applied for the orbital radii extending from negative to positive infinity.

Figure 1.6 shows plots of Eqs. (1.7-1) and (1.7-2) expressing respectively the toryx law of motion and the Kepler's third law of planetary motion. The highlights of these plots are:

- The toryx law of planetary motion is described by Eq. (1.7-1). It is applied to a range of b_1 extending from negative to positive infinity; within a range of b_1 extending from 0.5 to positive infinity the values of β_1 are expressed with real numbers, while within the remaining range of b_1 with imaginary numbers.

- The Kepler's third law of planetary motion is described by Eq. (1.7-2). It is applied to the range of b_1 extending from zero to positive infinity with all values of β_1 expressed with real numbers.

- When $b_1 > 5$, the difference between the values of β_1 calculated based on the toryx law of motion and the Kepler's third law of planetary motion becomes small and it decrease as b_1 increases. As b_1 decreases from 5 to 2, this difference progressively increases.

- According to the Kepler's third law of planetary motion, as b_1 decreases from 2 to 0, β_1 sharply increases and approaches positive infinity $(+\infty)$. According to the toryx law of motion, as b_1 decreases from 2 to 0. The value of β_1 initially slightly increases and then, after reaching its maximum value of 1 at $b_1 = 1$, it sharply decreases.

2. Features of Abstract Mathematics of a Toryx

Equations describing the toryx spacetime parameters are mostly based on elementary math commonly taught in high schools. However, to satisfy the toryx spacetime postulates, it is necessary to modify several aspects of elementary math, including the definitions of zero, number line and elementary trigonometric functions. Also, unlike the elementary math that deals with stationary spiral elements, the toryx math considers the spiral elements in motion, explaining its name.

2.1 Infinility versus Elementary Zero

Conventionally, we use the elementary zero (0) in two ways. Firstly, we use it for counting of non-divisible entities. In an elementary number line (Fig. 2.1) it appears as an integer immediately preceding number one (1).

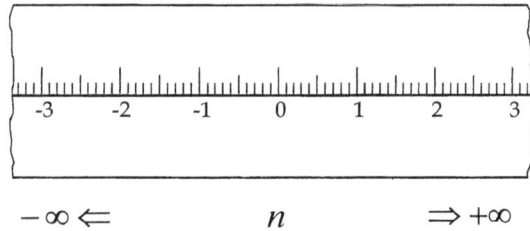

Figure 2.1.1. Elementary number line.

Secondly, we use zero to represent the absolute absence of any quantity and quality. Mathematically, the elementary zero (0) is equal to a ratio of one (1) to infinity (∞). The toryx math clearly separates two applications of zero described above. The zero is still considered as an integer for counting of non-divisible entities and still retains its old symbol (0). But, in application to the spacetime entities the zero is replaced with a quantity that is infinitely approaching to it. This quantity is called *infinility*, from the "infinite nil." (Notably, the term infinility is used in the toryx math instead of the known math term *infinitesimal*). In the toryx math, both infinity and infinility can be positive, negative, real and imaginary as shown below.

$$\text{Real infinility: } \pm 0 = \frac{1}{\pm \infty}; \quad \text{Imaginary infinility: } \pm 0i = \frac{1}{\pm \infty i}$$

$$\text{Real infinity: } \pm \infty = \frac{1}{\pm 0}; \quad \text{Imaginary infinity: } \pm \infty i = \frac{1}{\pm 0i}$$

Figure 2.1.2 shows symbolically positive and negative infinities $(\pm\infty)$ and also positive and negative infinility (± 0) as equal counterparts in respect to the positive and negative unities (± 1).

Infinility $\Leftarrow$ $\qquad\qquad\qquad$ $\Rightarrow$ *Infinity*

$\quad(\pm 0)$ $\qquad\qquad\qquad\quad$ $(\pm\vec{1})$ $\qquad\qquad\qquad$ $(\pm\infty)$

$\qquad\qquad\qquad\qquad\qquad$ *Unity*

Figure 2.1.2. Infinity $(\pm\infty)$, infinility (± 0) and unity (± 1).

2.2 Toryx Trigonometry

Definitions of elementary trigonometric functions are based on transformations of a right triangle as a function of the non-right angle φ_2 (Fig. 2.2.1):

$$\cos\varphi_2 = x \quad (0^0 < \varphi_2 < 360^0) \tag{2.2-1}$$

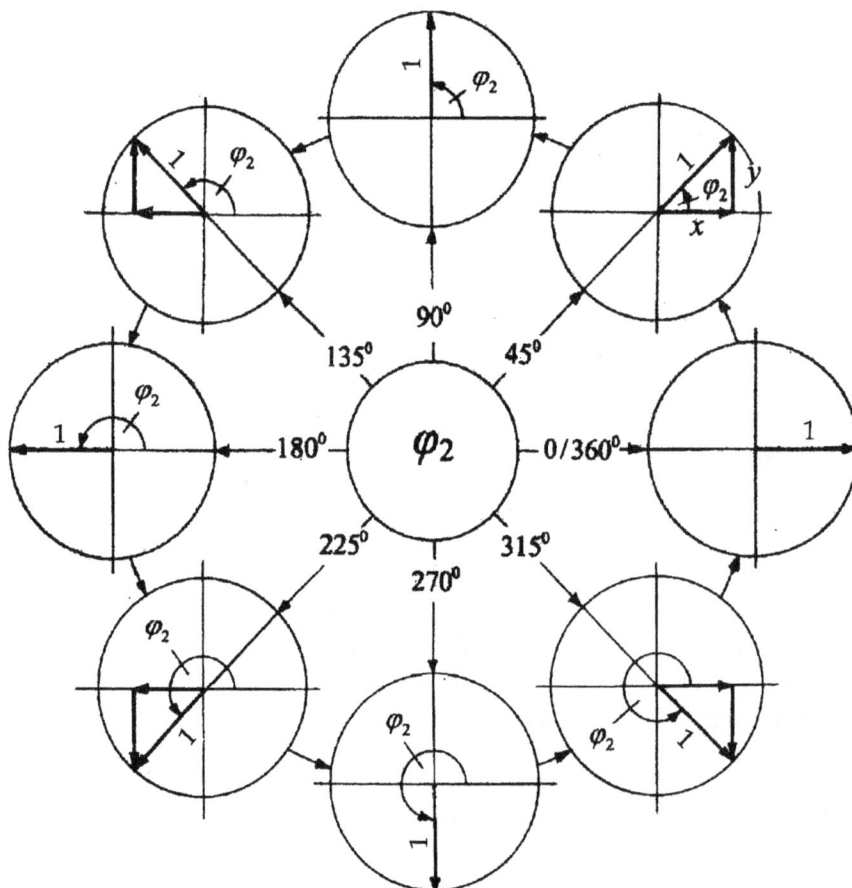

Figure 2.2.1. Transformations of a right triangle in elementary trigonometry.

The main features of the transformations shown in Fig. 2.2.1 are:

- When the length of the hypotenuse of the triangles is equal to 1, the ranges of the lengths of its sides x and y are between 1 and -1.
- The triangles located in two left quadrants are the mirror images of the triangles located in two right quadrants.
- The triangles located in two bottom quadrants are the mirror images of the triangles located at two top quadrants.

In the toryx trigonometry, the transformations of the right triangle are partially modified to satisfy the toryx spacetime postulates. Consequently, the toryx trigonometric function $\cos s\varphi_2$ relates to the elementary trigonometric function $\cos\varphi_2$ as follows:

$$\cos s\varphi_2 = \cos\varphi_2 \quad (0 < \varphi_2 < 180^0) \tag{2.2-2}$$

$$\cos s\varphi_2 = 1/\cos\varphi_2 \quad (180^0 < \varphi_2 < 360^0) \tag{2.2-3}$$

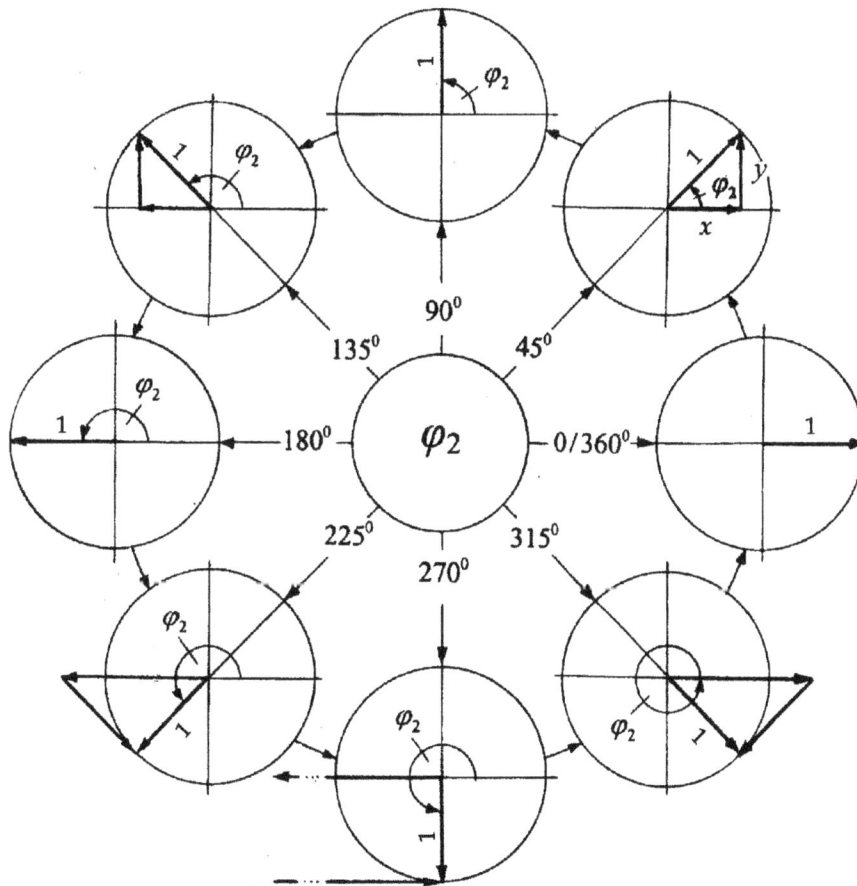

Figure 2.2.2. Transformations of a right triangle
in the toryx trigonometry.

The main features of the transformations shown in Fig. 2.2.2 are:

- When the angle φ_2 is between 0 and 180^0, the right triangles are the same as in elementary trigonometry. Thus, within this range of the angle φ_2 the elementary and toryx trigonometry are based on the same principle.
- When the angle φ_2 is between 180 and 360^0, the right triangle becomes **outverted**. Consequently, the length of its horizontal side x becomes greater than 1, while the length of the other side y is expressed with imaginary numbers.
- When the angle φ_2 approaches 270^0 from the angle smaller than 270^0, the length of its horizontal side x approaches real positive infinity $(+\infty)$, while the length of the other side y approaches imaginary positive infinity $(+\infty i)$.
- When the angle φ_2 approaches 270^0 from the angle greater than 270^0, the length of its horizontal side x approaches real negative infinity $(-\infty)$, while the length of the other side y approaches imaginary negative infinity $(-\infty i)$.
- When the angle φ_2 approaches 360^0 from the angle smaller than 360^0, the length of its horizontal side x approaches 1, while the length of the other imaginary side y approaches imaginary negative infinility $(-0i)$.

2.3 Toryx Number Lines

We consider below four kinds of toryx number lines that are directly related to the toryx parameters:

- *Toryx vorticity V number line*
- *Toryx reality R number line*
- *Toryx boundary B number line*
- *Toryx golden spacetime intensity S number line*
- *Toryx string period ratio P number line.*

All five number lines are presented in the forms of circular diagrams in which the numbers V, R, B, S and P are expressed as functions of the steepness angle of toryx trailing string φ_2.

Toryx vorticity V number line - In the toryx vorticity V number line (Fig. 2.3.1), the real numbers V are equal to the ratio of toryx trailing string radius r_2 to the toryx leading string radius r_1 with an opposite sign. These numbers are extended clockwise along a circle from the real positive infinity $(+\infty)$ to the real negative infinity $(-\infty)$ as a function of the steepness angle of trailing string φ_2.

$$V = -\frac{r_2}{r_1} = -\cos s\varphi_2 \qquad (2.3\text{-}1)$$

The toryx vorticity V number line is divided into two domains, the V ***infinility domain*** and the V ***infinity domain***, occupying equal sectors of the circular number line.

- The V infinility domain occupies two top quadrants; it contains the values of V extending clockwise from the real positive unity $(+1)$ and passing through infinility (± 0) to the real negative unity (-1).

- The V infinity domain resides in two bottom quadrants; it contains the values of V extending counterclockwise from the real positive unity $(+1)$ and passing through infinity $(\pm\infty)$ to the real negative unity (-1).

In the toryx vorticity V number line, the real positive infinility $(+0)$ merges with real negative infinility (-0) at $\varphi_2 = 90^0$, while real negative infinity $(-\infty)$ merges with real positive infinity $(+\infty)$ at $\varphi_2 = 270^0$.

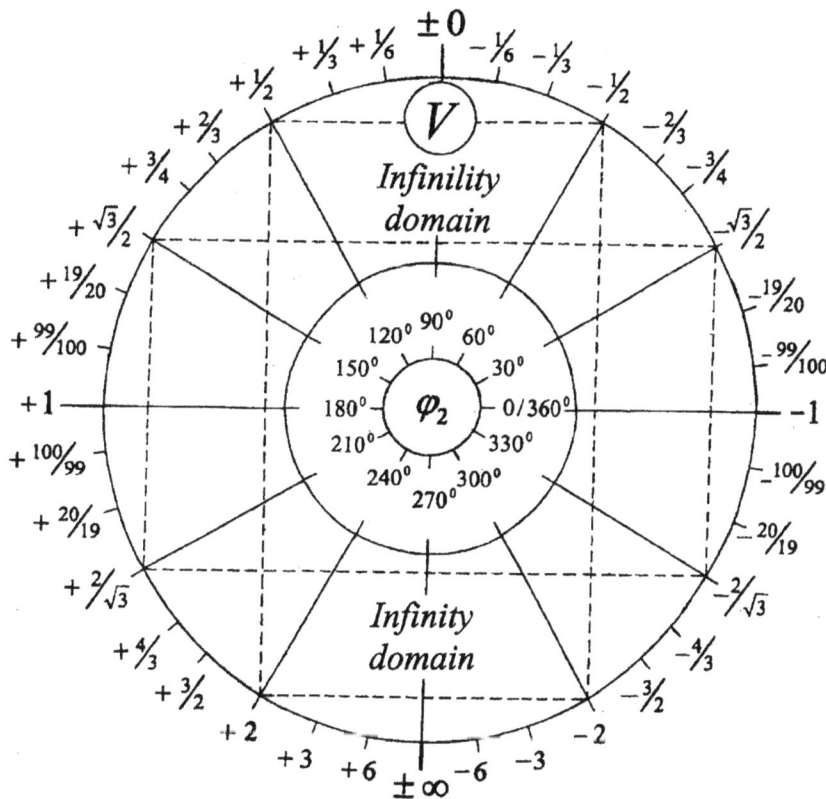

Figure 2.3.1. The toryx vorticity V number line.

There are two kinds of symmetries between the numbers V that belong to the four quadrants of circular diagram, the ***inverse V-symmetry*** and the ***reverse V-symmetry***.

- In the inverse V-symmetry, the magnitudes of the numbers V located in the top quadrants are inversed (reciprocated) in respect to the magnitudes of the numbers V located in the bottom quadrants.

- In the reverse V-symmetry, the numbers V located in the right quadrants and the left quadrants have the same magnitudes but reversed signs.

Toryx reality R number line - In the toryx reality R number line (Fig. 2.3.2), the real and imaginary numbers R are equal to the ratio of toryx trailing string wavelength λ_2 to the circular length of toryx eye $l_0 = 2\pi r_0$. These numbers are extended counterclockwise along a circle from the real positive infinity $(+\infty)$ to the imaginary negative infinity $(-\infty i)$ as a function of the steepness angle of trailing string φ_2.

$$R = \frac{\lambda_2}{2\pi r_0} = \sqrt{\frac{1 + \cos s\varphi_2}{1 - \cos s\varphi_2}} \qquad (2.3\text{-}2)$$

The toryx reality R number line is divided into two domains, the **R infinility domain** and the **R infinity domain**, occupying equal sectors of the circular number line.

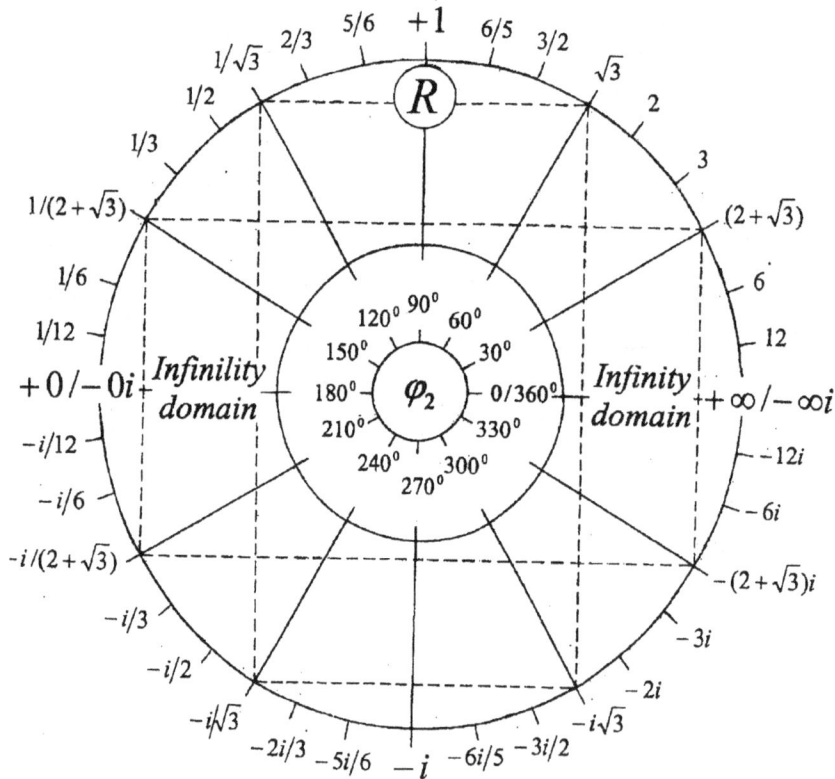

Figure 2.3.2. The toryx reality R number line.

- <u>The R infinility domain</u> occupies two left quadrants; it contains the values of R extending counterclockwise from the real positive unity $(+1)$ and passing through infinility $(+0/0i)$ to the imaginary negative unity $(-i)$.

- The R infinity domain resides in two right quadrants; it contains the values of R extending clockwise from the real positive unity $(+1)$ and passing through real positive and imaginary negative infinities $(+\infty/-\infty i)$ to the imaginary negative unity $(-i)$.

In the toryx reality R number line, the real positive infinility $(+0)$ merges with the imaginary negative infinility $(-0i)$ at $\varphi_2 = 180^0$, while real positive infinity $(+\infty)$ merges with imaginary negative infinity $(-\infty i)$ at $\varphi_2 = 360^0$.

There are two kinds of symmetries between the numbers R that belong to the four quadrants of circular diagram, the ***inverse R-symmetry*** and the ***reverse reality R-symmetry***.

- In the inverse R-symmetry, the magnitudes of the numbers R located in the left quadrants are inversed (reciprocated) in respect to the magnitudes of the numbers R located in the right quadrants.
- In the reverse reality R-symmetry, the numbers R located in the top quadrants are real positive, while these numbers in the bottom quadrants are imaginary negative.

Toryx boundary B number line - In the toryx boundary B number line (Fig. 2.3.3), the real numbers B are equal to the ratio of the radius of toryx radius of spherical boundary r to the toryx eye radius r_0.

$$B = \frac{r}{r_0} = \frac{1 + \cos s\varphi_2}{1 - \cos s\varphi_2} \tag{2.3-3}$$

These numbers are extended counterclockwise along a circle from the real positive infinity $(+\infty)$ to the real negative infinity $(-\infty)$ as a function of the steepness angle of trailing string φ_2.

The toryx boundary B number line is divided into two domains, the ***B infinility domain*** and the ***B infinity domain***, occupying equal sectors of the circular number line.

- The B infinility domain occupies two left quadrants; it contains the values of B extending counterclockwise from the real positive unity $(+1)$ and passing through infinility $(+0/-0)$ to the real negative unity (-1).
- The B infinity domain resides in two right quadrants; it contains the values of B extending clockwise from the real positive unity $(+1)$ and passing through real positive and negative infinities $(+\infty/-\infty)$ to the real negative unity (-1).

In the toryx boundary B number line, real positive infinility $(+0)$ merges with real negative infinility (-0) at $\varphi_2 = 180^0$, while real positive infinity $(+\infty)$ merges with real negative infinity $(-\infty)$ at $\varphi_2 = 360^0$.

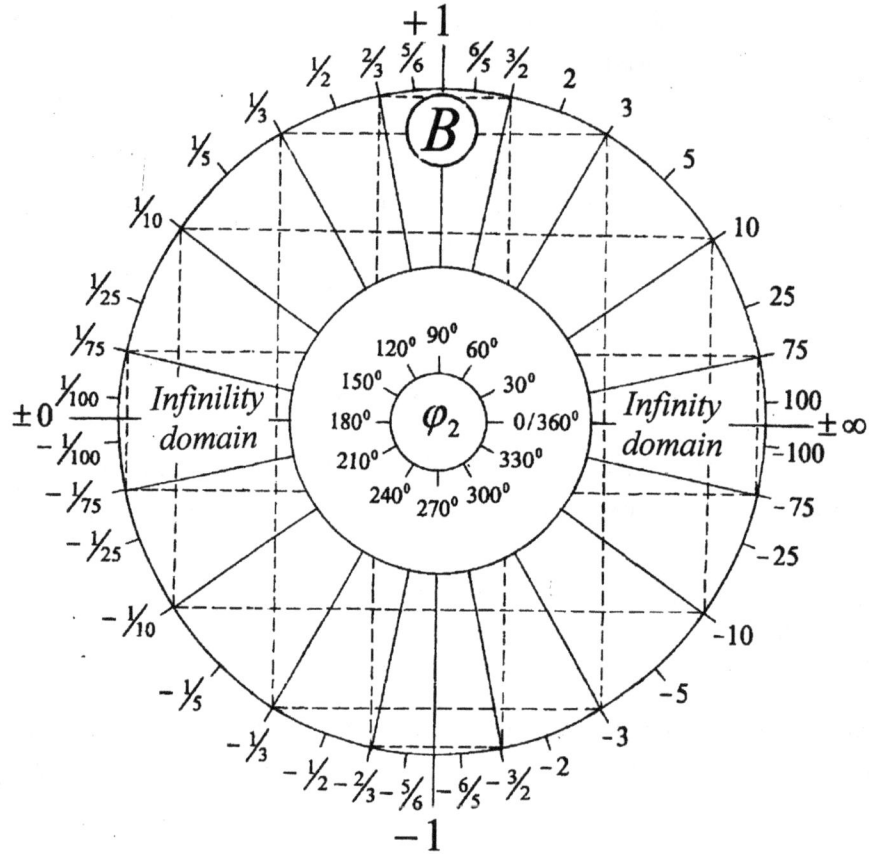

Figure 2.3.3. The toryx boundary B number line.

There are two kinds of symmetries between the numbers B that belong to the four quadrants of circular diagram, the *inverse B-symmetry* and the *reverse B-symmetry*.

- In the inverse B-symmetry, the magnitudes of the numbers B located in the left quadrants are inversed (reciprocated) in respect to the magnitudes of the numbers B located in the right quadrants.
- In the reverse B-symmetry, the numbers B located in the top and bottom quadrants have the same magnitudes but reversed signs.

Toryx golden spacetime intensity S number line – In the toryx golden spacetime intensity S number line (Fig. 2.3.4), the numbers S are related to the toryx radius of spherical boundary r, the eye radius r_0, the radius of leading string r_1 and the radius of trailing string r_2 of a toryx by the equation:

$$S = -\frac{r\, r_0}{r_1 r_2} = -\frac{1 - \cos s^2 \varphi_2}{\cos s \varphi_2} \qquad (2.3\text{-}4)$$

When $S = \pm 1$, $\cos s \varphi_2$ relates to the *golden ratio* $\phi = 1.618033989$ as shown in Table 2.3.

Table 2.3. $\cos s\varphi_2$ versus golden ratio ϕ.

Steepness angle of trailing string φ_2	$\cos s\varphi_2$	S
51.83^0	$+1/\phi$	-1
128.17^0	$-1/\phi$	$+1$
231.83^0	$-1/\phi$	-1
308.17^0	$+1/\phi$	$+1$

The toryx golden spacetime intensity S number line is divided into four domains, two **S-infinity domains** and two **S-infinility domains**.

- The S-infinity domains occupy top and bottom equal segments of the circular diagram. In both top and bottom S-infinity domains, the values of S extend clockwise from the real positive unity $(+1)$ and passing through infinity $(\pm\infty)$ to the real negative unity (-1).
- The S-infinility domains occupy left and right equal segments of the circular diagram. In both left and right S-infinility domains, the values of S extend clockwise from the real negative unity (-1) and passing through infinility (± 0) to the real positive unity $(+1)$.

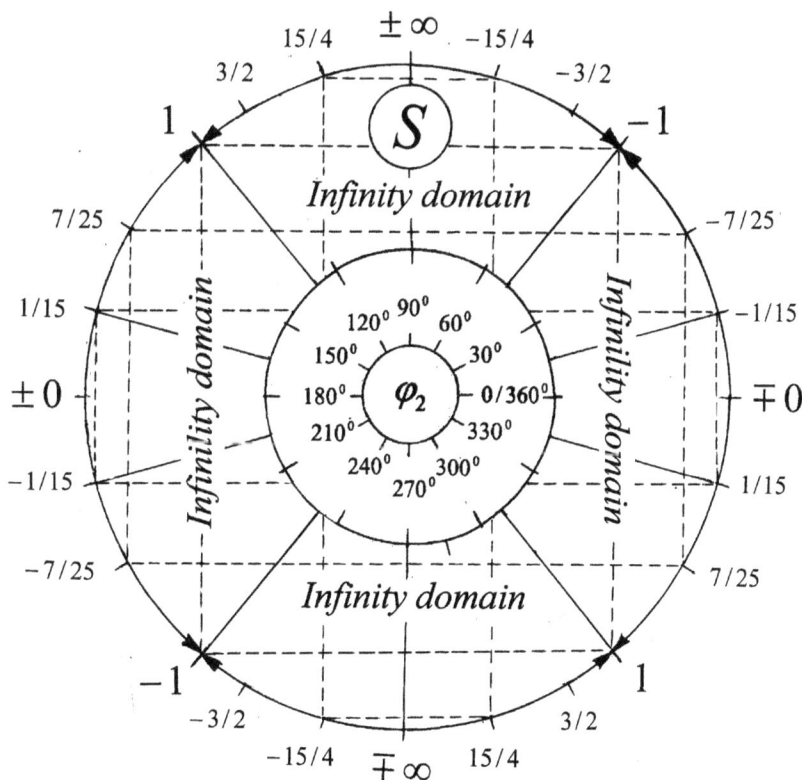

Figure 2.3.4. The toryx golden spacetime intensity S number line.

The numbers located in each of four quadrants of the S number line relate to the numbers of adjacent quadrants according to the ***reverse S-symmetry***. It means that the numbers S located <u>in</u> <u>each of four quadrants</u> have the same magnitudes but reversed signs in respect to the numbers S located in their adjacent quadrants.

Toryx string period ratio P number line – In the toryx string period ratio P number line (Fig. 2.3.5), the numbers P are related to the ratio of the periods of toryx trailing string T_2 and leading string T_1 by the equation:

$$P = \frac{T_2}{T_1} = \sin s\varphi_2 \tag{2.3-5}$$

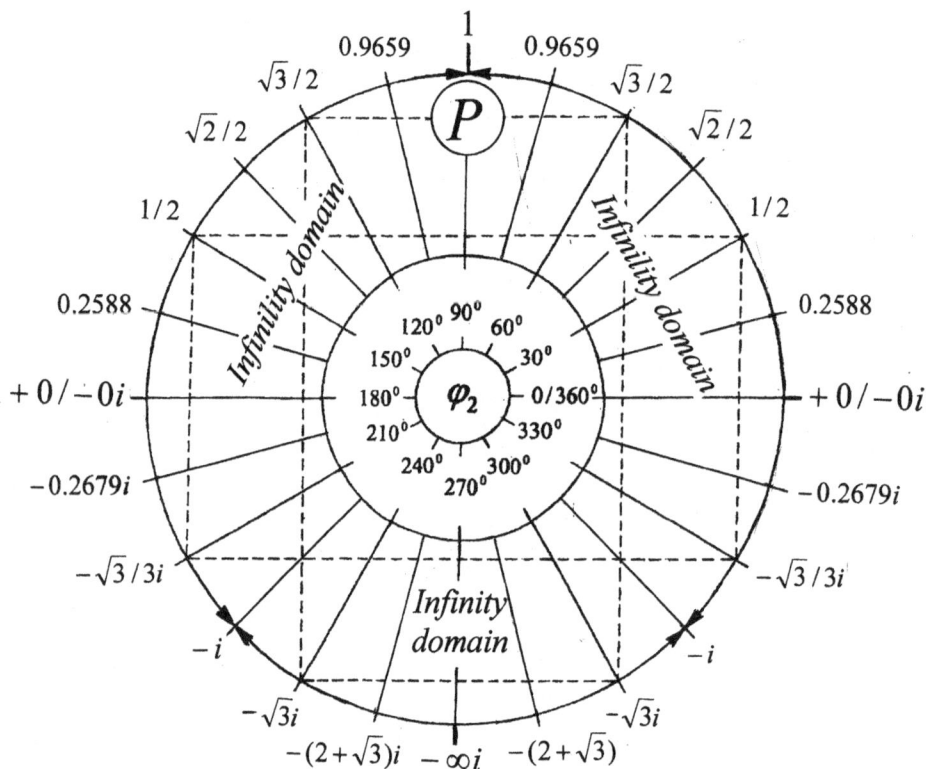

Figure 2.3.5. The toryx string period ratio P number line.

The toryx string period ratio P number line is divided into three domains, two ***P-infinility domains*** and one ***P-infinity domains***.

- <u>Two *P*-infinility domains</u> occupy top left and right equal segments of the circular diagram. In both left and right top segments, the values of P extend respectively counter-clockwise and clockwise from the real positive unity $(+1)$ and passing through infinility $(+0/-0i)$ to the imaginary negative unity $(-i)$. The numbers P located in the left and right top segments are equal to one another.

- The *P*-infinity domain occupy bottom left and right segments of the circular diagram. both left and right bottom segments, the values of P extend from the imaginary negative unity $(-i)$ and pass through imaginary negative infinity $(-\infty i)$ to the imaginary negative unity $(-i)$. The numbers P located in the left and right bottom segments are equal to one another.

2.4. Toryx Parameters in Number Lines & Trigonometry

The numbers V of the toryx vorticity V number line (Fig. 2.3.1) relate to the toryx parameters by the equation:

$$V = -\frac{r_2}{r_1} = -\frac{b_2}{b_1} = -\frac{b_1 - 1}{b_1} = -\frac{b-1}{b+1} = -\beta_{2r} = -\cos s\varphi_2 \qquad (2.4\text{-}1)$$

The numbers R of the toryx reality R number line (Fig. 2.3.2) relate to the toryx parameters by the equation:

$$R = \eta_2 = \sqrt{b} = \sqrt{2b_1 - 1} = b_1\beta_{2t} = \sqrt{\frac{1 + \cos s\varphi_2}{1 - \cos s\varphi_2}} \qquad (2.4\text{-}2)$$

The numbers B of the toryx boundary B number line (Fig. 2.3.3) relate to the toryx parameters by the equation:

$$B = b = \frac{r}{r_0} = 2b_1 - 1 = \frac{1 + \cos s\varphi_2}{1 - \cos s\varphi_2} \qquad (2.4\text{-}3)$$

The numbers S of the toryx golden spacetime intensity S number line (Fig. 2.3.4) are related to the toryx parameters by the equation:

$$S = -\frac{r\, r_0}{r_1\, r_2} = -\frac{b}{b_1\, b_2} = -\frac{4b}{b^2 - 1} = -\frac{2b_1 - 1}{b_1(b_1 - 1)} = -\frac{\beta_{2t}^2}{\beta_{2r}} = -\frac{1 - \cos s^2\varphi_2}{\cos s\varphi_2} \qquad (2.4\text{-}4)$$

When $S = \pm 1$, the toryx parameters relate to the golden ratio ϕ as shown in Table 2.4.

Table 2.4. Relationships between toryx parameters and
the golden ratio ϕ when $S = \pm 1$.

φ_2	S	V	b_2	b_1	b
51.83^0	-1	$-1/\phi$	ϕ	$1 + \phi$	$1 + 2\phi$
128.17^0	$+1$	$+1/\phi$	$-1/\phi^2$	$1/\phi$	$2/\phi - 1$
231.83^0	-1	$+1/\phi$	$-1/\phi$	$1/\phi^2$	$2/\phi^2 - 1$
308.17^0	$+1$	$-1/\phi$	$-(1 + \phi)$	$-\phi$	$-(1 + 2\phi)$

The numbers P of the toryx string period ratio P number line relate to the toryx parameters by the equation:

$$P = \frac{T_2}{T_1} == \frac{t_2}{t_1} = \frac{\sqrt{2b_1 - 1}}{b_1} = \frac{1}{w_2} = \beta_1 = \beta_{1r} = \beta_{2t} = \sin s\varphi_2 \qquad (2.4\text{-}5)$$

Figure 2.4 shows the application of the spiral spacetime math for the calculation of the relative velocities of the toryx trailing string corresponding to the middle point of the trailing string as its steepness angle φ_2 increases from 0 to 360^0. In each right triangle of velocities of trailing string one side represents the relative translational velocity β_{2t} and the other side the relative rotational velocity β_{2r}, while its hypotenuse represents the relative spiral velocity $\beta_2 = 1$.

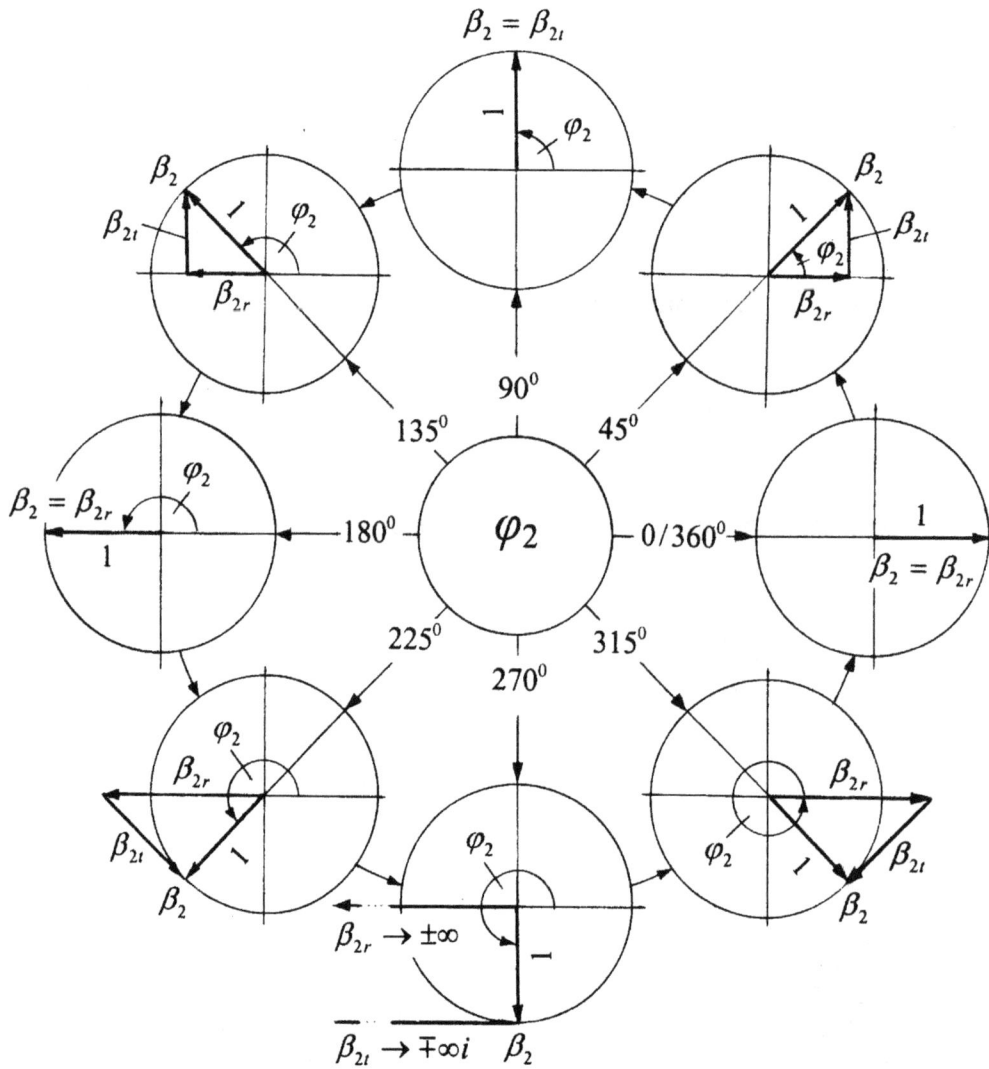

Figure 2.4. Transformations of right triangle representing vectors of the relative velocities of toryx trailing string β_2, β_{2t} and β_{2r} corresponding to the middle point a of trailing string (Fig. 1.1.3).

Notes:

<u>Notes:</u>

3. CLASSIFICATION OF TORYCES

CONTENTS

3.1 Groups of Toryces

Depending on the toryx vorticity V and reality R numbers the toryces are divided into four *main groups* as shown in Figure 3.1 and Tables 3.1.1 and 3.1.2.

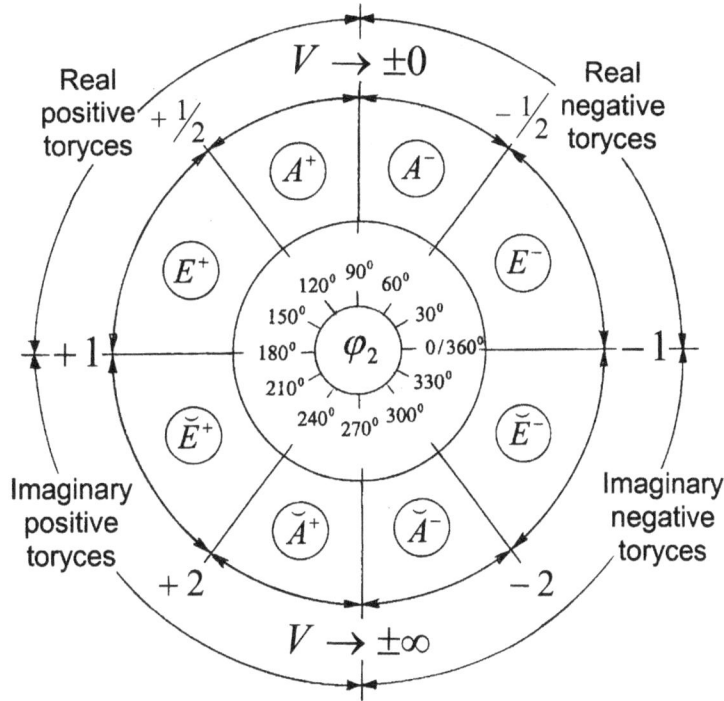

Figure 3.1. Main groups and subgroups of toryces.

Real negative toryces $(0^0 < \varphi_2 < 90^0)$ – The real negative toryces are located in the top right quadrants of the circular diagrams. In the mid of trailing strings of these toryces, both the translational velocity β_{2t} and the rotational velocity β_{2r} are *subluminal* and expressed with *real numbers*.

Real positive toryces $(90^0 < \varphi_2 < 180^0)$ – The real positive toryces are located in the top left quadrants of the circular diagrams. In the mid of trailing strings of these toryces, both the translational velocity β_{2t} and the rotational velocity β_{2r} are *subluminal* and expressed with *real numbers*.

Imaginary positive toryces $(180^0 < \varphi_2 < 270^0)$ – The imaginary positive toryces are located in the bottom left quadrants of the circular diagrams. In the mid of trailing strings of these toryces, the rotational velocities β_{2r} are *superluminal* and expressed with *real numbers*, whereas the translational velocities β_{2t} are expressed with *imaginary numbers*.

Imaginary negative toryces $(270^0 < \varphi_2 < 360^0)$ – The imaginary negative toryces are located in the bottom right quadrants of the circular diagrams. In the mid of trailing strings of these toryces, the rotational velocities β_{2r} are *superluminal* and expressed with *real numbers*, whereas the translational velocities β_{2t} are expressed with *imaginary numbers*.

Table 3.1.1. The realities R, the vorticities V and the mid relative rotational velocities β_{2r} of trailing strings of toryces of main groups.

Toryx name	φ_2	R	V	β_{2r}
Real negative	0^0 - 90^0	Real	(–)	< 1 (subluminal)
Real positive	90^0 - 180^0	Real	(+)	< 1 (subluminal)
Imaginary positive	180^0 - 270^0	Imaginary	(+)	> 1 (superluminal)
Imaginary negative	270^0 - 360^0	Imaginary	(–)	>1 (superluminal)

Notably, superluminal and subluminal components of spiral velocity of toryx trailing string are respectively associated with absorption and release of spacetime by the toryx.

Within each main group, the toryces are further divided into two *subgroups* as shown in Fig. 3.1 and Table 3.1.2.

Table 3.1.2. Toryces of main groups and subgroups.

Toryces of main groups		Toryces of subgroups	
Name	Symbols	Name	φ_2
Real negative	$A^- $ & E^-	E^-	$0^0 < \varphi_2 < 60^0$
		A^-	$60^0 < \varphi_2 < 90^0$
Real positive	$A^+ $ & E^+	A^+	$90^0 < \varphi_2 < 120^0$
		E^+	$120^0 < \varphi_2 < 180^0$
Imaginary positive	$\breve{A}^+ $ & $\breve{E}^+$	$\breve{E}^+$	$180^0 < \varphi_2 < 240^0$
		$\breve{A}^+$	$240^0 < \varphi_2 < 270^0$
Imaginary negative	$\breve{A}^- $ & $\breve{E}^-$	$\breve{A}^-$	$270^0 < \varphi_2 < 300^0$
		$\breve{E}^-$	$300^0 < \varphi_2 < 360^0$

3.2 Vorticities & Realities of Adjacent Toryces

Main groups of toryces - The vorticities V and the realities R of toryces of adjacent main groups are symmetrically related as shown in Figs 2.3.1, 2.3.2 and Table 3.2.1.

Table 3.2.1. Symmetrical relationships between the vorticity V and the reality R of polarized toryces of adjacent main groups.

Toryces of main groups		Eqs. (a)	Eqs. (b)
Reality-polarized negative toryces	$\breve{E}^- \, \& \, E^-$ Eq. (3.2-1)	$\breve{V}_E^- = 1/V_E^-$	$\breve{R}_E^- = \pm i R_E^-$
Reality-polarized positive toryces	$\breve{E}^+ \, \& \, E^+$ Eq. (3.2-2)	$\breve{V}_E^+ = 1/V_E^+$	$\breve{R}_E^+ = \pm i R_E^+$
Vorticity-polarized real toryces	$A^+ \, \& \, A^-$ Eq. (3.2-3)	$V_A^+ = -V_A^-$	$R_A^+ = 1/R_A^-$
Vorticity-polarized imaginary toryces	$\breve{A}^+ \, \& \, \breve{A}^-$ Eq. (3.2-4)	$\breve{V}_A^+ = -\breve{V}_A^-$	$\breve{R}_A^+ = 1/\breve{R}_A^-$

Table 3.2.2 shows the relationships between the relative radii of leading strings b_1 and spherical boundary b of adjacent vorticity- and reality-polarized toryces of main groups.

Table 3.2.2. Relationships between the relative radii of leading string b_1 and spherical boundary b of adjacent polarized toryces of main groups.

Toryces of main groups		Eqs. (a)	Eqs. (b)
Reality-polarized negative toryces	$\breve{E}^- \, \& \, E^-$ Eq. (3.2-5)	$\breve{b}_{1E}^- = 1 - b_{1E}^-$	$\breve{b}_E^- = -b_E^-$
Reality-polarized positive toryces	$\breve{E}^+ \, \& \, E^+$ Eq. (3.2-6)	$\breve{b}_{1E}^+ = 1 \quad b_{1E}^+$	$\breve{b}_E^+ = -b_E^+$
Vorticity-polarized real toryces	$A^+ \, \& \, A^-$ Eq. (3.2-7)	$b_{1A}^+ = \dfrac{b_{1A}^-}{2b_{1A}^- - 1}$	$b_A^+ = \dfrac{1}{b_A^-}$
Vorticity-polarized imaginary toryces	$\breve{A}^+ \, \& \, \breve{A}^-$ Eq. (3.2-8)	$\breve{b}_{1A}^+ = \dfrac{\breve{b}_{1A}^-}{2\breve{b}_{1A}^- - 1}$	$\breve{b}_A^+ = \dfrac{1}{\breve{b}_A^-}$

Subgroups groups of toryces - Exhibit 3.2 shows a limitation of degrees of freedom for adjacent real negative toryces $A^- \, \& \, E^-$ at the borderline between them. Table 3.2.3 shows the relationships between vorticities of toryces of subgroups.

Exhibit 3.2. Vorticities of adjacent real negative A^- & E^- toryces at the borderline between them.

The vorticities of adjacent real negative toryces A^- & E^- at the borderline between them are the same and they are equal to:

$$V_A^- = V_E^- = -0.5$$

(3.2-9)

Table 3.2.3. Relationships between vorticities of toryces of subgroups.

Toryces of subgroups		Equations
Real negative toryces	A^- & E^-	$V_A^- + V_E^- = -1$ (3.2-10)
Real positive toryces	A^+ & E^+	$V_A^+ + V_E^+ = +1$ (3.2-11)
Imaginary positive toryces	$\breve{A}^+$ & $\breve{E}^+$	$\dfrac{1}{V_A^+} + \dfrac{1}{V_E^+} = +1$ (3.2-12)
Imaginary negative toryces	$\breve{A}^-$ & $\breve{E}^-$	$\dfrac{1}{V_A^-} + \dfrac{1}{V_E^-} = -1$ (3.2-13)

Table 3.2.4 summarizes the relationships between the relative radii of leading string b_1 and spherical boundary b of toryces of subgroups.

Table 3.2.4. Relationships between the relative radii of leading string b_1 and spherical boundary b of toryces of subgroups.

Toryces of subgroups		Eqs. (a)	Eqs. (b)
Negative toryces	A^- & E^- Eq. (3.2-14)	$b_{1A}^- = \dfrac{b_{1E}^-}{b_{1E}^- - 1}$	$b_A^- = \dfrac{b_E^- + 3}{b_E^- - 1}$
Positive toryces	A^+ & E^+ Eq. (3.2-16)	$b_{1A}^+ = \dfrac{b_{1E}^+}{3b_{1E}^+ - 1}$	$b_A^+ = \dfrac{1 - b_E^+}{1 + 3b_E^+}$

3.3 Self-Polarized Toryces

The self-polarized toryces is a special case of the real vorticity-polarized toryces. In Section 3.1, we defined main groups of toryces based on their reality R and vorticity V. In the real toryces of the main groups, all toryx parameters related to the middle point a of trailing string (Fig. 1.1.3) are expressed with real numbers, while in the imaginary toryces of the main groups some toryx parameters are expressed with imaginary numbers.

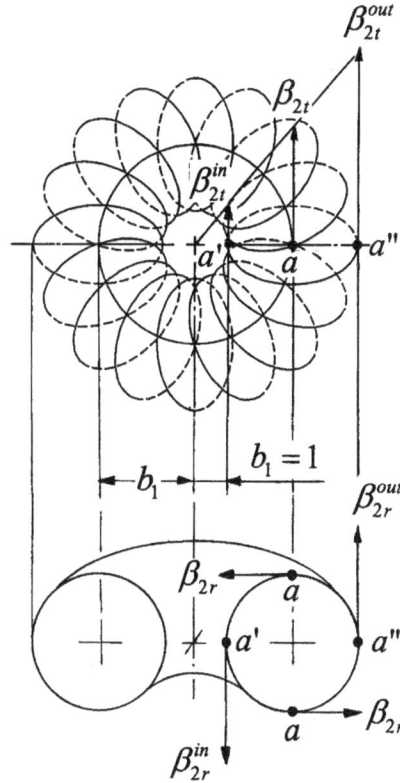

Figure 3.3.1. Relative peripheral velocities of trailing string.

Notably, this is true only when the translational velocity β_{2t}, the rotational velocity β_{2r}, the wavelength η_2, and the steepness φ_2 correspond to the middle point a of toryx trailing string (Fig. 1.1.3). These parameters vary during each cycle of trailing string as illustrated in Fig. 3.3.1 in application to the translational velocity β_{2t} and the rotational velocity β_{2r} of toryx trailing string, creating a condition for the existence the so-called ***self-polarized toryces*** in which the realities change from real to imaginary during each cycle of their trailing string.

Table 3.3 shows equations for relative peripheral translational and rotational velocities of trailing strings.

Table 3.3. Relative peripheral velocities of trailing string.

Velocities At the inner point a' (Fig. 3.3.1)		Velocities at the outer point a'' (Fig. 3.3.1)	
$\beta_{2t}^{in} = \dfrac{\sqrt{2b_1-1}}{b_1^2}$	(3.3-1)	$\beta_{2t}^{out} = \dfrac{(2b_1-1)^{3/2}}{b_1^2}$	(3.3-2)
$\beta_{2r}^{in} = \dfrac{\sqrt{b_1^4-2b_1+1}}{b_1^2}$	(3.3-3)	$\beta_{2r}^{out} = \dfrac{\sqrt{b_1^4-(2b_1-1)^3}}{b_1^2}$	(3.3-4)

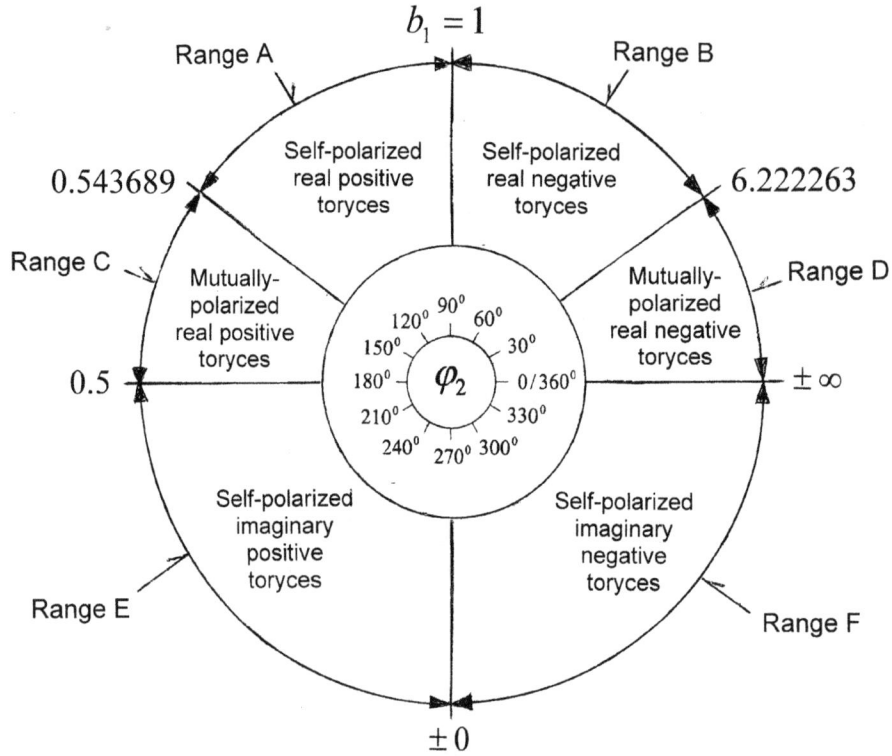

Figure 3.3.2. The ranges of the relative radii of leading strings b_1 of self-polarized and mutually-polarized toryces.

At each point of trailing string, the translational velocity must be proportional to the distance of said point from the toryx center. At the same time, the rotational velocity of trailing string must vary to satisfy Eq. (1.5-3) according to which the trailing string must propagate at a constant relative spiral string velocity $\beta_2 = 1$. From Eqs. (3.3-1) – (3.3-4) we find the following values of the inner and outer translational and rotational velocities of trailing strings β_{2t}^{in}, β_{2r}^{in}, β_{2t}^{out} and β_{2r}^{out} within the ranges A through F of the relative radii of toryx leading strings b_1 (Fig. 3.3.2):

- **Range A** $(0.543689 > b_1 > 1)$ ***Self-polarized real positive toryces*** - In these toryces, the values of the inner translational velocities β_{2t}^{in} are real and superluminal, while the values of the inner rotational velocities β_{2r}^{in} are imaginary and subluminal. The relative inner translational velocity β_{2t}^{in} reaches its maximum of 1.299038 at $b_1 = \frac{2}{3}$ $(\varphi_2 = 120^0)$.

- **Range B** $(1 > b_1 > 0.6222263)$ ***Self-polarized real negative toryces*** - In these toryces, the values of the outer translational velocities β_{2t}^{out} are real and superluminal, while the values of outer rotational velocities β_{2r}^{out} are imaginary and subluminal,. The relative outer translational velocity β_{2t}^{out} reaches its maximum of 1.299038 at $b_1 = 2$ $(\varphi_2 = 60^0)$.

- **Range C** $(0.5 > b_1 > 0.543689)$ *Mutually-polarized real positive toryces* - In these toryces, the values of both the inner and outer rotational velocities β_{2r}^{in} and β_{2r}^{out} as well as the inner and outer translational velocities β_{2t}^{in} and β_{2t}^{out} are real and subluminal.

- **Range D** $(0.6222263 > b_1 > +\infty)$ *Mutually-polarized real negative toryces* - In these toryces, the values of both the inner and outer rotational velocities β_{2r}^{in} and β_{2r}^{out} as well as the inner and outer translational velocities β_{2t}^{in} and β_{2t}^{out} are real and subluminal.

- **Range E** $(0.5 > b_1 > +0)$ *Self-polarized imaginary positive toryces* - In these toryces, the values of both the inner and outer rotational velocities β_{2r}^{in} and β_{2r}^{out} are real and superluminal, while the values of both the inner and outer translational velocities β_{2t}^{in} and β_{2t}^{out} are imaginary, but either subluminal or superluminal.

- **Range F** $(-0 > b_1 > -\infty)$ *Self-polarized imaginary negative toryces* - In these toryces, the values of both the inner and outer rotational velocities β_{2r}^{in} and β_{2r}^{out} are real and superluminal, while the values of both the inner and outer translational velocities β_{2t}^{in} and β_{2t}^{out} are imaginary, but either subluminal or superluminal.

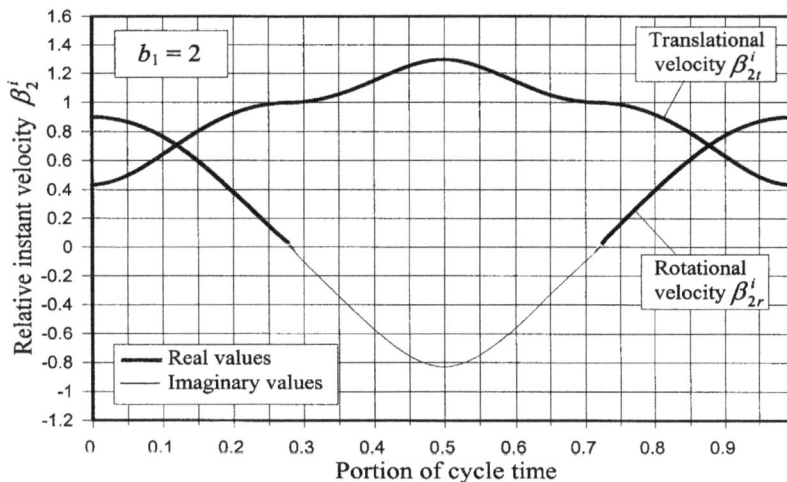

Figure 3.3.3. Variation of instant velocities β_{2t}^{i} and β_{2r}^{i} of trailing string within one cycle of trailing string.

Figure 3.3.3 shows the variation of the relative instant translational and rotational velocities, β_{2t}^{i} and β_{2r}^{i} for the case when $b_1 = 2$. Notably, between 0.25 and 0.75 portions of the cycle of trailing string, β_{2t}^{i} exceeds velocity of light, while β_{2r}^{i} is expressed with imaginary numbers.

<u>Notes</u>

4. TRENDS OF TORYX PARAMETERS

CONTENTS

The toryces having the spacetime parameters described in the previous Chapters are called the **basic toryces**. Presented below are the plots of equations expressing the basic toryx parameters as functions of the steepness angle of trailing string φ_2.

4.1 Radii of Leading & Trailing Strings

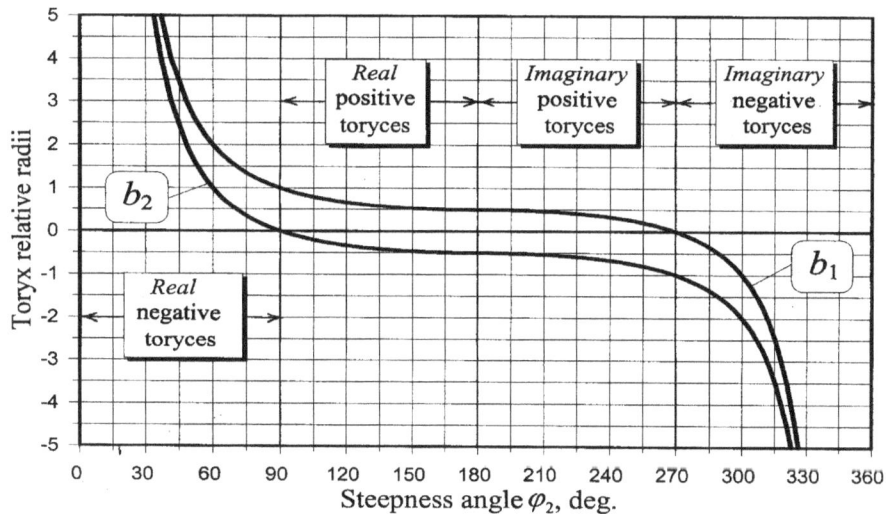

φ_2	$360/0^0$	90^0	180^0	270^0
b_1	$-\infty/+\infty$	$+1$	$+\frac{1}{2}$	$+0/-0$
b_2	$-\infty/+\infty$	$+0/-0$	$-\frac{1}{2}$	-1

Figure 4.1. Trends of the relative radii of leading and trailing strings, b_1 and b_2.

4.2 Radius of Spherical Boundary

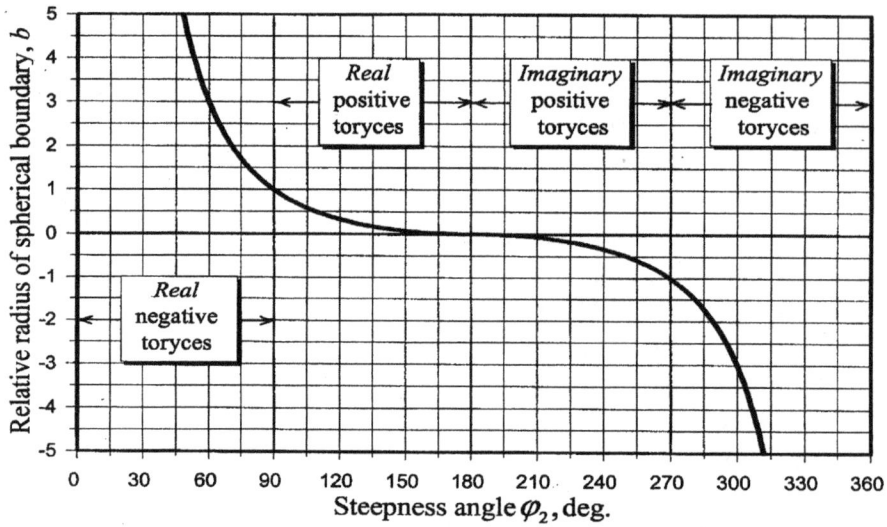

φ_2	$360/0^0$	90^0	180^0	270^0
b	$-\infty/+\infty$	$+1$	$+0/-0$	-1

Figure 4.2. Trend of the toryx relative radius of spherical boundary b.

4.3 Wavelength of Trailing String

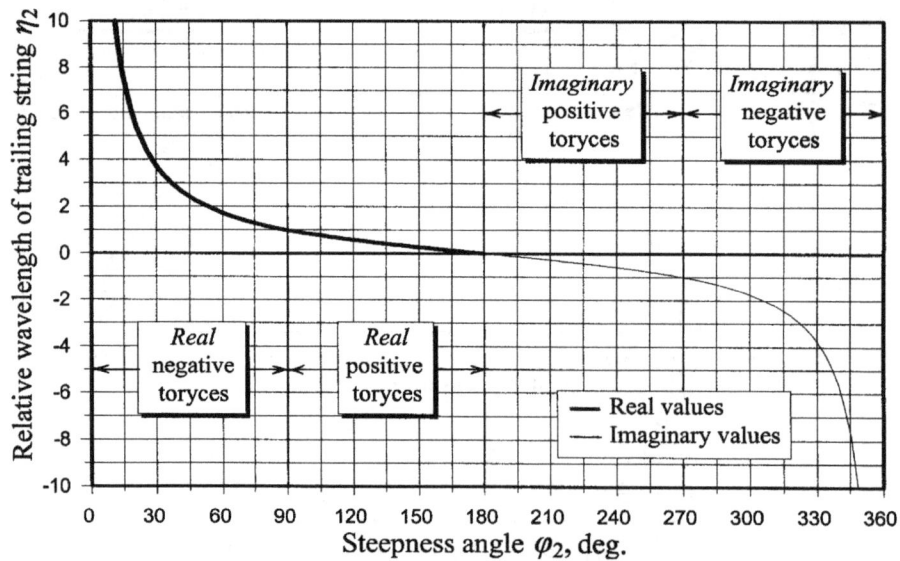

φ_2	$360/0^0$	90^0	180^0	270^0
η_2	$-\infty i/+\infty$	$+1$	$+0/-0i$	$-i$

Figure 4.3. Trend of the relative wavelength of trailing string η_2.

4.4 The Number of Windings of Trailing String

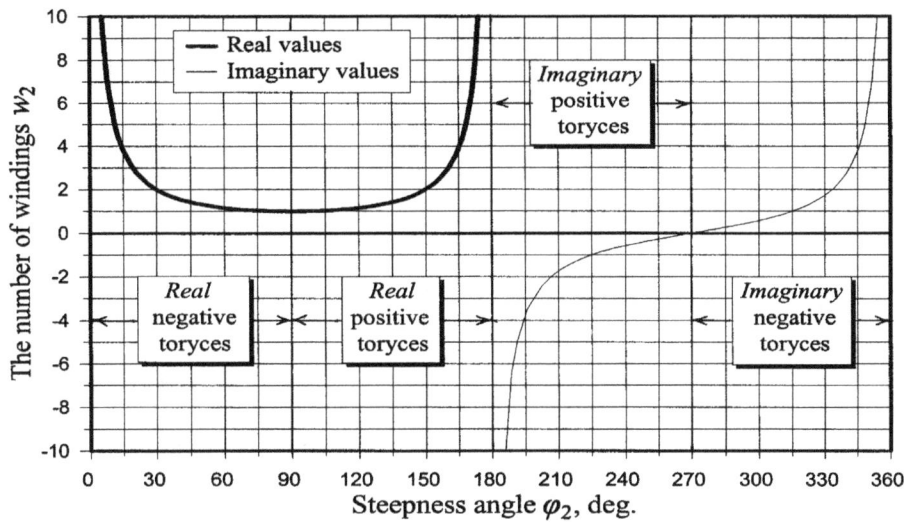

φ_2	$360/0^0$	90^0	180^0	270^0
w_2	$+\infty i/+\infty$	$+1$	$+\infty/-\infty i$	$-0i/+0i$

Figure 4.4. Trend of the number of windings of trailing string w_2.

4.5 Translational Velocity of Trailing String

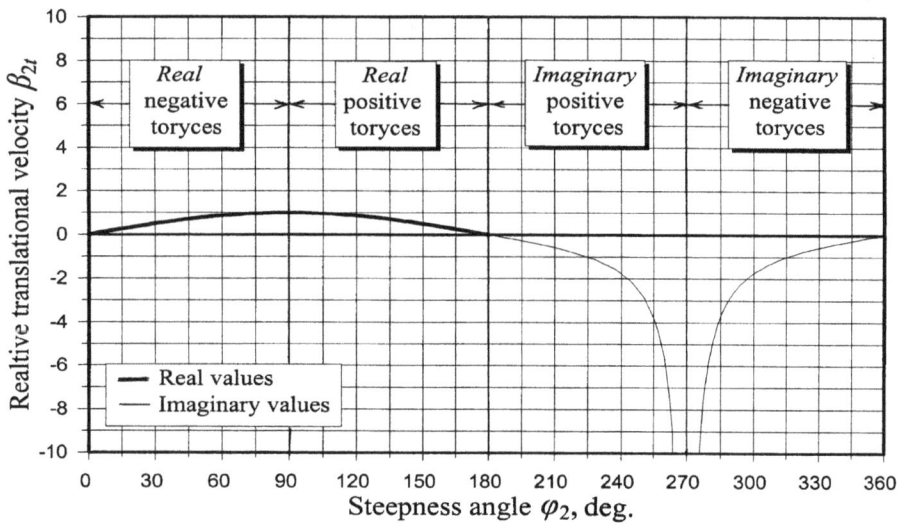

φ_2	$360/0^0$	90^0	180^0	270^0
β_{2t}	$-0i/+0$	$+1$	$+0/-0i$	$-\infty i$

Figure 4.5. Trend of the relative translational velocity of trailing string β_{2t}.

4.6 Rotational Velocity of Trailing String

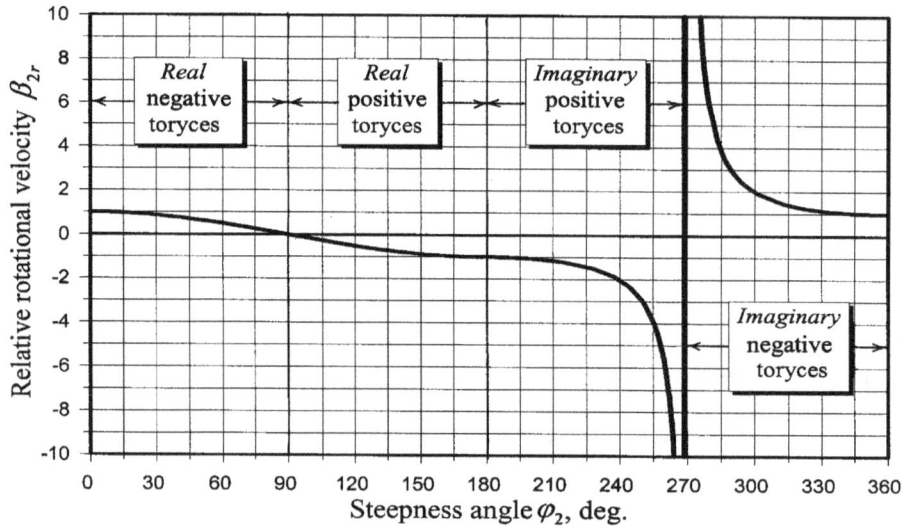

φ_2	$360/0^0$	90^0	180^0	270^0
β_{2r}	$+1$	$+0/-0$	-1	$-\infty/+\infty$

Figure 4.6. Trend of the relative rotational velocity of trailing string β_{2r}.

4.7 Frequency of Leading String

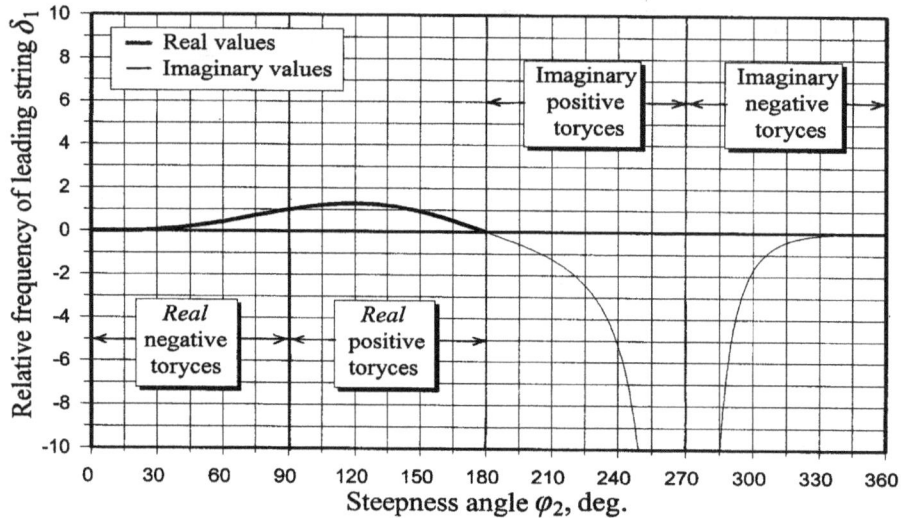

φ_2	$360/0^0$	90^0	180^0	270^0
δ_1	$-0i/+0$	$+1$	$+0/-0i$	$-\infty i$

Figure 4.7. Trend of the relative frequency of leading string δ_1.

4.8 Frequency of Trailing String

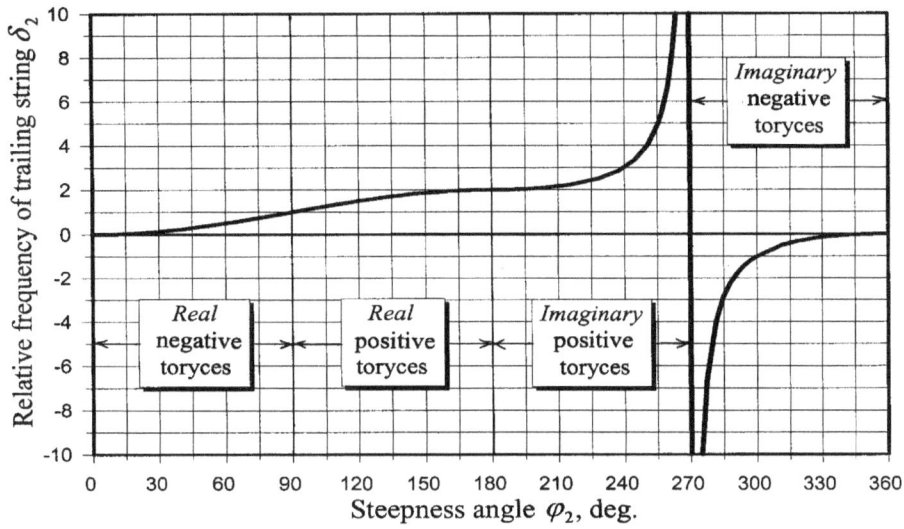

φ_2	$360/0^0$	90^0	180^0	270^0
δ_2	$-0/+0$	$+1$	$+2$	$+\infty/-\infty$

Figure 4.8. Trend of the relative frequency of trailing string δ_2.

4.9 Period of Leading String

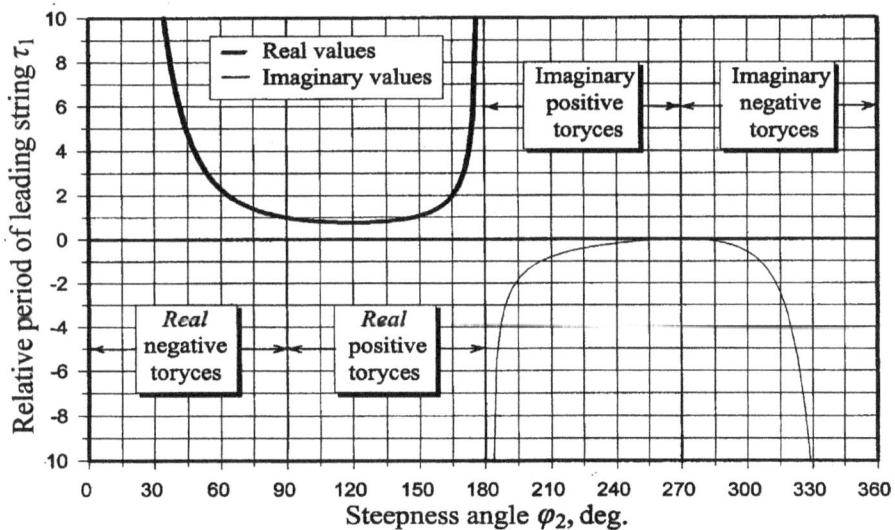

φ_2	$360/0^0$	90^0	180^0	270^0
τ_1	$-\infty i/+\infty$	$+1$	$+\infty/-\infty i$	$-0i$

Figure 4.9. Trend of the relative period of leading string τ_1.

4.10 Period of Trailing String

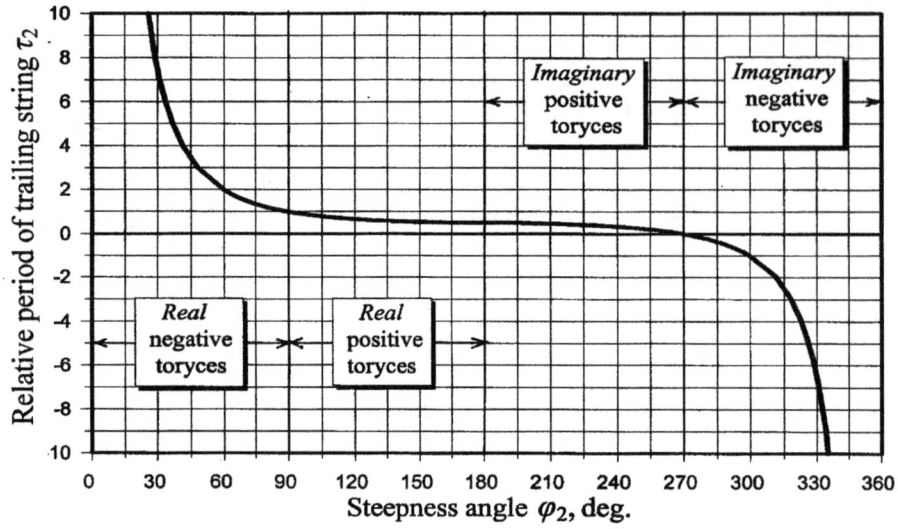

φ_2	$360/0^0$	90^0	180^0	270^0
τ_2	$-\infty/+\infty$	$+1$	$+\frac{1}{2}$	$+0/-0$

Figure 4.10. Trend of the relative period of trailing string τ_2.

5. INVERSION STATES OF TORYCES

5.1 Inversion of Toryx Leading & Trailing Strings

As the radius of toryx leading string b_1 decreases from positive to negative infinity, the leading string and trailing string undergo through spacetime transformations that can be visualized as described below.

Inversion of leading string – Visualize the toryx leading string in the form of an extremely thin and narrow circular ribbon with the relative radius b_1 (Fig. 5.1.1). When $b_1 > 0$; the outer color of the circular ribbon is assumed to be black while its inner color is white. The leading string remains **outverted** until b_1 reduces to positive infinility $(+0)$. When b_1 approaches infinility (± 0), the leading string becomes **inverted**. So, when $b_1 < 0$, the outer color of leading string appears white, while its inner color looks black.

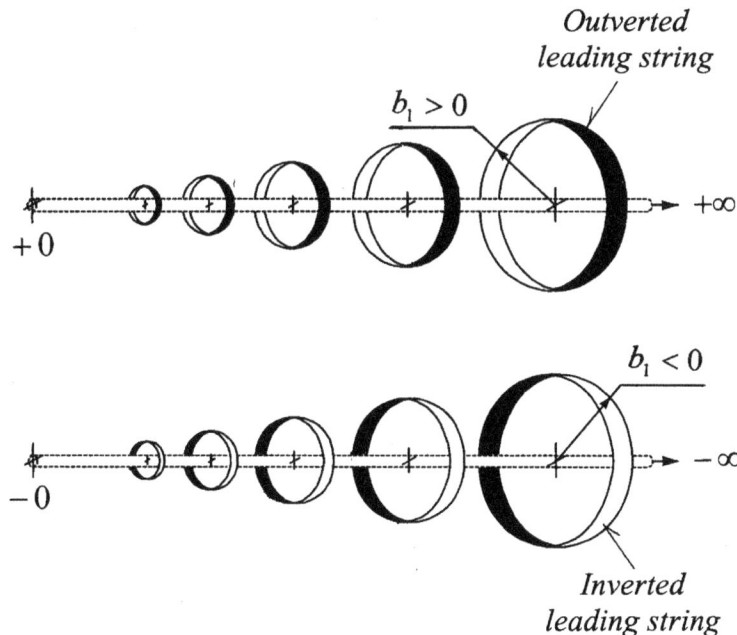

Fig. 5.1.1. Inversion of a leading string.

Inversion of a trailing string - Visualize the toryx trailing string in the form of an extremely thin and narrow toroidal ribbon with the relative radius b_2 (Fig. 5.1.2). When the relative radius of leading string $b_1 > 1$ the radius of trailing string $b_2 > 1$. For that case, the outer color of toroidal ribbon is assumed to be black, while its inner color white. The trailing string remains **outverted** until b_2 reduces to positive infinity $(+0)$. When $b_1 = 1$, b_2 approaches infinility (±0) and the trailing string becomes **inverted**. So, when $b_2 < 1$, the outer color of leading string appears white, while its inner color looks black.

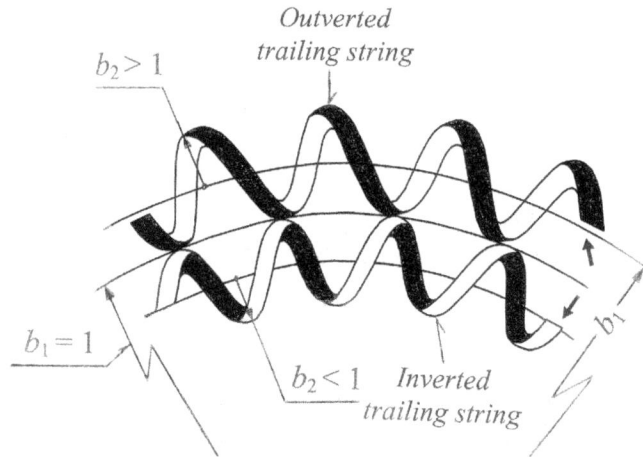

Fig. 5.1.2. Inversion of a trailing string.

Inversion of a spherical boundary - Visualize an extremely thin spherical boundary with the relative radius b (Fig. 5.1.3).

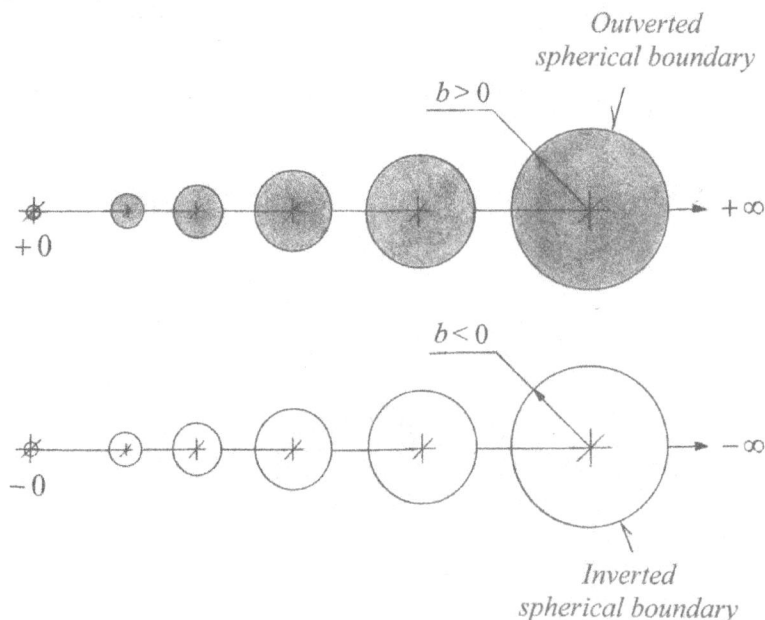

Fig. 5.1.3. Inversion of a spherical boundary.

When $b > 0$, the outer color of the spherical boundary is assumed to be grey and its inner color is assumed to be white. The spherical boundary remains *outverted* until b reduces to positive infinility $(+0)$. When b approaches infinility (± 0), the spherical boundary becomes *inverted*. So, when $b < 0$, the outer color of spherical boundary appears white, while its inner color becomes grey.

5.2 Inversion of Toryces

Figure 5.2 shows that as the radius of the toryx leading string decreases from positive to negative infinity, the steepness angle of trailing string φ_2 increases from 0^0 to 360^0, while the *toryx vorticity* V extends equally from *Unity* (± 1) towards both *Infinility* (± 0) and *Infinity* $(\pm \infty)$.

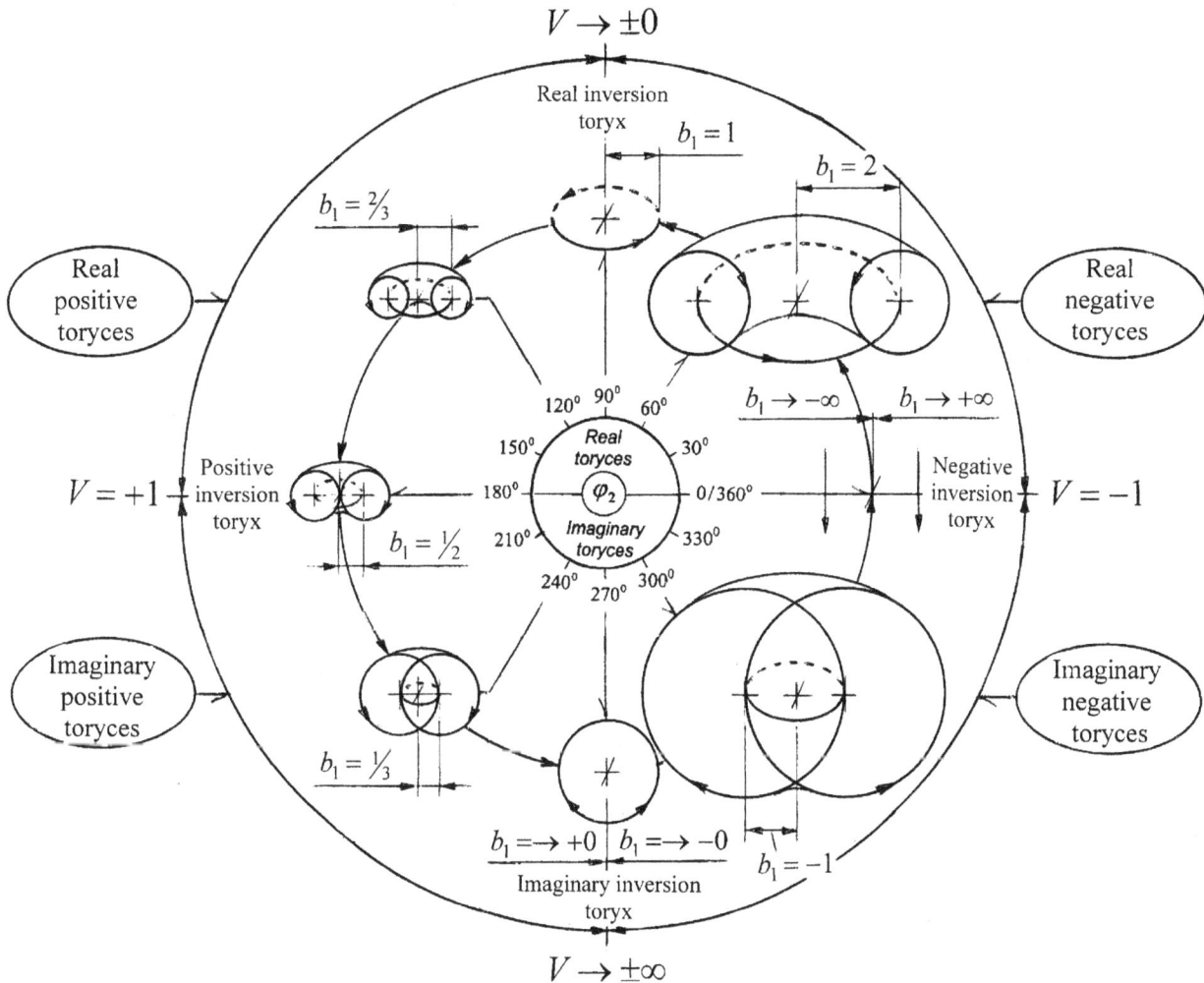

Figure 5.2. Metamorphoses of toryx leading and trailing strings as a function of the steepness angle of trailing string φ_2.

Four kinds of *inversion toryces* are located at the boundaries of four quadrants of the circular diagram.

- **Negative inversion toryx** $(\varphi_2 \to 0^0/360^0)$ - At this point, the toryx leading string, trailing string and its wavelength become inverted, and the toryx appears as two parallel lines separated by the distance equal to the diameter of real inversion string.

- **Real inversion toryx** $(\varphi_2 \to 90^0)$ – At this point, the toryx trailing string becomes inverted $(b_1 \to +1, b_2 \to \pm0, \eta_2 = 1)$, and the toryx appears as a circle with the relative radius $b_1 \to +1$.

- **Positive inversion toryx** $(\varphi_2 \to 180^0)$ – At this point, the wavelength of toryx trailing string becomes inverted $(b_1 \to +0.5, b_2 \to -0.5, \eta_2 \to +0/-0i)$, and the toryx appears as an extreme case of a *spindle torus* with the inner parts of its windings touching one another.

- **Imaginary inversion toryx** $(\varphi_2 \to 270^0)$ – At this point, the toryx leading string becomes inverted $(b_1 \to \pm0, b_2 \to -1, \eta_2 = -i)$, and the toryx appears as a circle with the relative radius approaching -1. The circle is located at the plane perpendicular to the plane of the real inversion string.

Table 5.2 shows extreme relative parameters of inversion toryces.

Table 5.2. Extreme relative parameters of inversion toryces.

Inversion toryces	φ_2	b	b_1	b_2	η_2	w_2	β_{2t}	β_{2r}	δ_1	δ_2
Negative inversion toryx	$0^0/360^0$	$-\infty$ $+\infty$	$-\infty$ $+\infty$	$-\infty$ $+\infty$	$-\infty i$ $+\infty$	$+\infty i$ $+\infty$	$-0i$ $+0$	$+0$ -0	$-0i$ $+0$	-0 $+0$
Real inversion toryx	90^0	-	-	$+0$ -0	-	-	-	$+0$ -0	-	-
Positive inversion toryx	180^0	$+0$ -0	$+0$ -0	-	$+0$ $-0i$	$+\infty$ $-\infty i$	$+0$ $-0i$	$-\infty$ $+\infty$	$+0$ $-0i$	-
Imaginary inversion toryx	270^0	-	-	-	-	$-0i$ $+0i$	$-\infty i$	-	$-\infty i$	$+\infty$ $-\infty$

5.3 Transformations of Real Negative Toryces

Real negative toryces (Fig. 5.3) belong to the top right quadrant of the circular diagram shown in Fig. 5.2. Trailing strings of these toryces are wound counter-clockwise outside of the real inversion string. As φ_2 increases, both b_1 and b_2 decrease, so that the trailing string appears like a conventional torus.

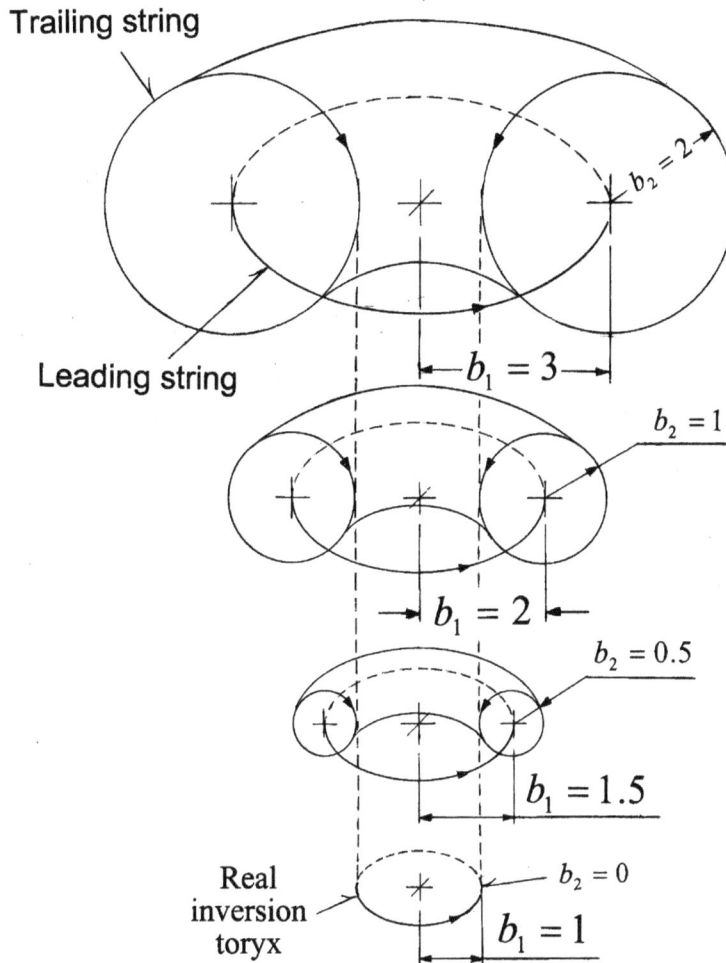

Range	φ_2	b	b_1	b_2	η_2	w_2	β_{2t}	β_{2r}
From	0^0	$+\infty$	$+\infty$	$+\infty$	$+\infty$	$+\infty$	$+0$	1.0
To	90^0	1.0	1.0	$+0$	1.0	1.0	1.0	$+0$

Figure 5.3. Transformations of real negative toryces.

5.4 Transformations of Real Positive Toryces

Real positive toryces (Fig. 5.4) belong to the top left quadrant of the circular diagram shown in Figure 5.2. Within this range the trailing string is inverted, so that its windings are now wound clockwise inside the real inversion toryx. As φ_2 increases, b_1 decreases, while the negative value of b_2 increases. Consequently, the toryx appears as an inverted toroidal spiral.

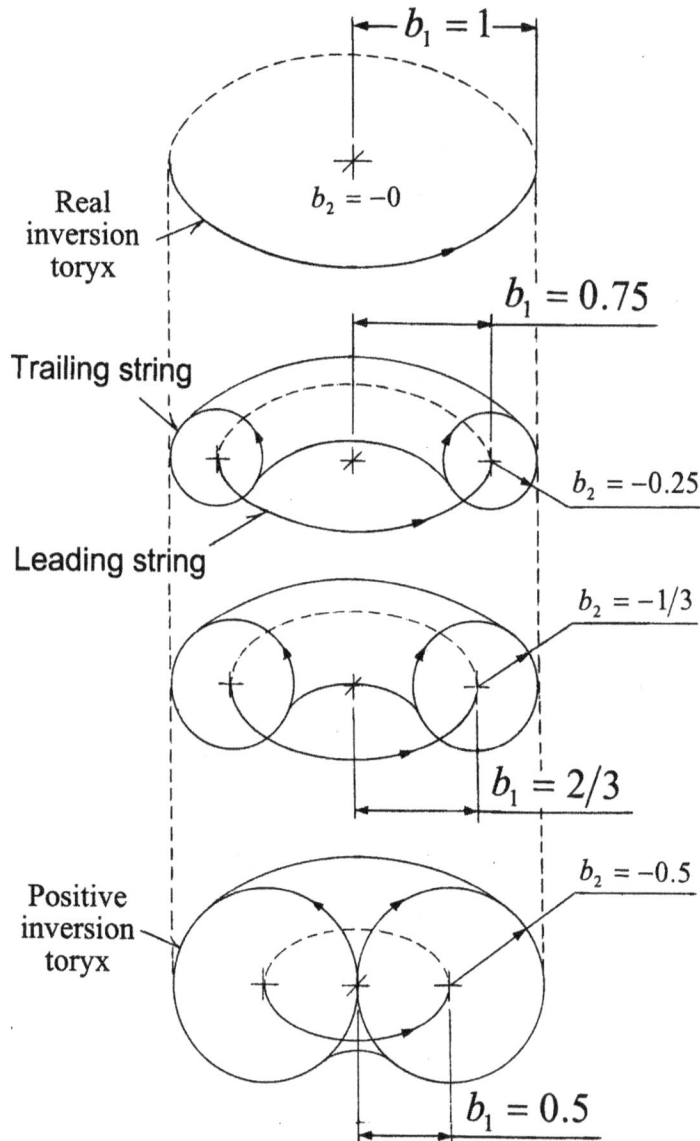

Range	φ_2	b	b_1	b_2	η_2	w_2	β_{2t}	β_{2r}
From	90^0	1.0	1.0	- 0	1.0	1.0	1.0	- 0
To	180^0	+0	0.5	-0.5	+0	$+\infty$	+0	-1.0

Figure 5.4. Transformations of real positive toryces.

5.5 Transformations of Imaginary Positive Toryces

Real positive toryces (Fig. 5.5) belong to the bottom left quadrant of the circular diagram shown in Figure 5.2. As φ_2 increases, b_1 decreases while negative values of b_2 increase. Within this range, the opposite parts of windings of trailing string intersect with one another like in a spindle torus.

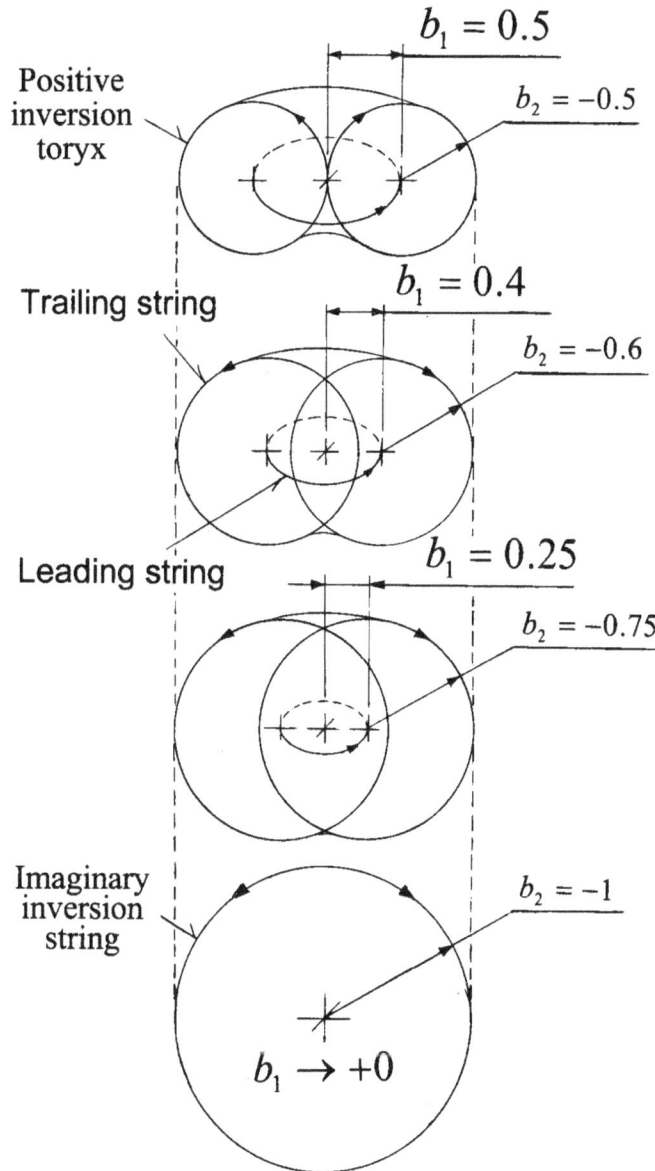

Range	φ_2	b	b_1	b_2	η_2	w_2	β_{2t}	β_{2r}
From	180^0	-0	0.5	-0.5	$-0i$	$-\infty i$	$-0i$	-1.0
To	270^0	-1.0	+0	-1.0	$-i$	$-0i$	$-\infty i$	$-\infty$

Figure 5.5. Transformations of imaginary positive toryces.

5.6 Transformations of Imaginary Negative Toryces

Imaginary negative toryces (Fig. 5.6) belong to the bottom right quadrant of the circular diagram shown in Figure 5.2. Here the leading string becomes inverted. As φ_2 increases, the negative values of b_1 increase, the negative values of b_2 also increase. Within this range the toryx windings are located outside of the imaginary inversion toryx.

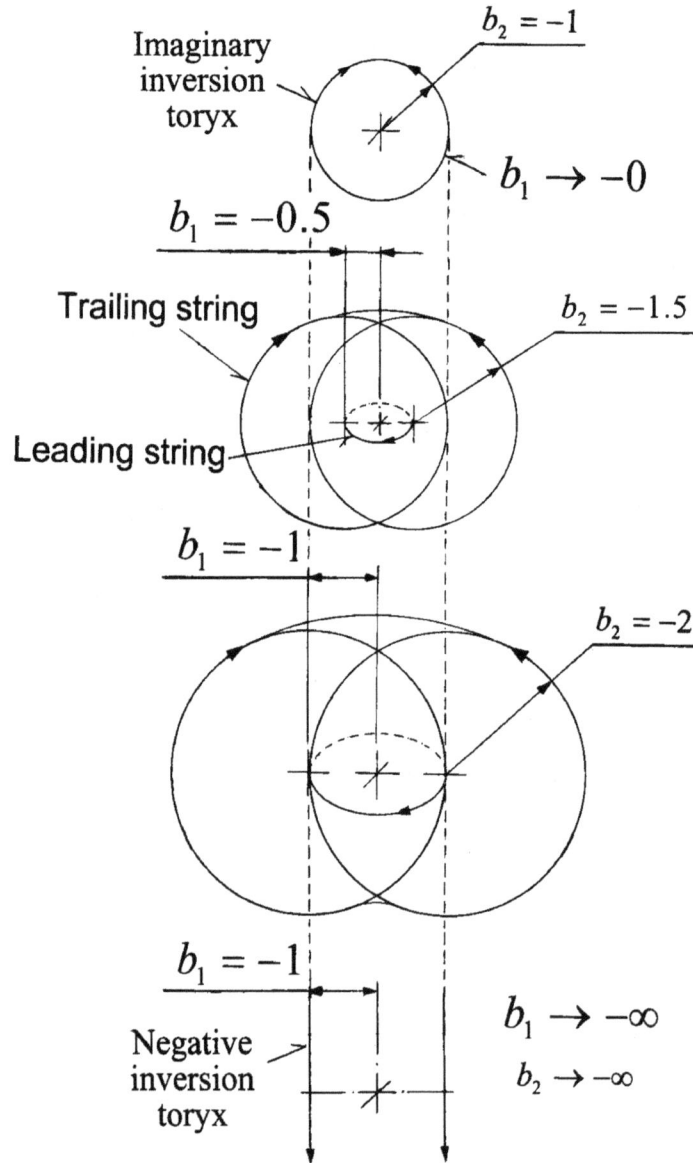

Range	φ_2	b	b_1	b_2	η_2	w_2	β_{2t}	β_{2r}
From	270^0	-1.0	-0	-1.0	$-i$	$+0i$	$-\infty i$	$+\infty$
To	360^0	$-\infty$	$-\infty$	$-\infty$	$-\infty i$	$+\infty i$	$-0i$	1.0

Figure 5.6. Transformations of imaginary negative toryces.

5.7 Summary of Toryx Transformations

Figure 5.7 summarizes transformation of toryces by showing together the metamorphoses of toryx spherical boundary, leading string and trailing string with the respective relative radii b, b_1 and b_2 as the steepness angle of trailing string φ_2 increases from 0^0 to 360^0.

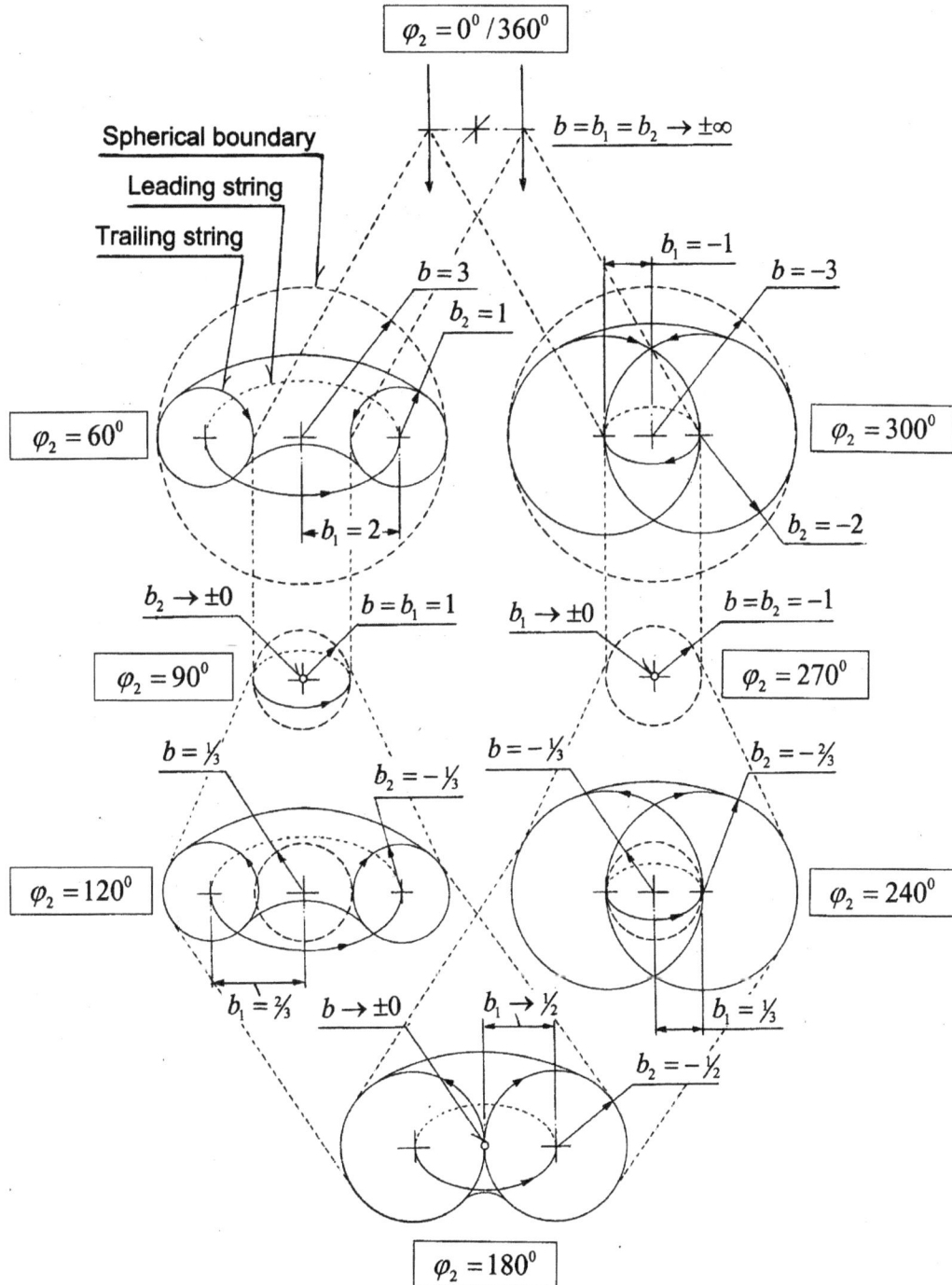

Figure 5.7. Transformations of toryx spherical boundary, leading string and trailing string as a function of the steepness angle of trailing string φ_2.

- When $\varphi_2 \to +0^0$, all three radii b, b_1 and b_2 approach positive infinity $(+\infty)$.

- Within the range of $\varphi_2 (+0^0 < \varphi_2 < 90^0)$, b, b_1 and b_2 decrease as φ_2 increases.

- When $\varphi_2 \to 90^0$, $b = b_1 = 1$ and b_2 approaches positive infinility $(+0)$; the ***trailing string becomes inverted*** and b_2 becomes negative.

- Within the range of $\varphi_2 (90^0 < \varphi_2 < 180^0)$, b_1 and b continue to decrease, while negative values of b_2 increase starting from negative infinility (-0).

- When $\varphi_2 \to 180^0$, $b_1 = \frac{1}{2}$, $b_2 = -\frac{1}{2}$ and b approaches positive infinility $(+0)$; the ***toryx spherical boundary becomes inverted*** and b becomes negative.

- Within the range of $\varphi_2 (180^0 < \varphi_2 < 270^0)$, b_1 continues to decrease, negative values of b_2 continue to increase, while negative values of b increase starting from negative infinility (-0).

- When $\varphi_2 \to 270^0$, $b = b_2 = -1$, while b_1 approaches positive infinility $(+0)$; the ***leading string becomes inverted*** and b_1 becomes negative.

- Within the range of $\varphi_2 (270^0 < \varphi_2 < 360^0)$, negative values of b_1 increase starting from negative infinility (-0), while negative values of b and b_2 continue to increase.

- When $\varphi_2 \to 360^0$, all three radii b, b_1 and b_2 approach negative infinity $(-\infty)$.

Table 5.7. Inversion points of toryx leading string, trailing strings and spherical boundary.

Toryx component	Relative radius	Steepness angle
Spherical boundary	$b \to \pm\infty$	$\varphi_2 \to 0^0/360^0$
	$b \to \pm 0$	$\varphi_2 \to 180^0$
Leading string	$b_1 \to \pm\infty$	$\varphi_2 \to 0^0/360^0$
	$b_1 \to \pm 0$	$\varphi_2 \to 270^0$
Trailing string	$b_2 \to \pm\infty$	$\varphi_2 \to 0^0/360^0$
	$b_2 \to \pm 0$	$\varphi_2 \to 90^0$

Table 5.7 summarizes the inversion steepness angles of trailing string and corresponding radii of toryx leading string, trailing string and spherical boundary.

5.8 Golden Toryces

Figure 5.8 and Table 2.4 show cross-sections and parameters of four golden toryces corresponding to the toryx golden polarization $G = \pm 1$.

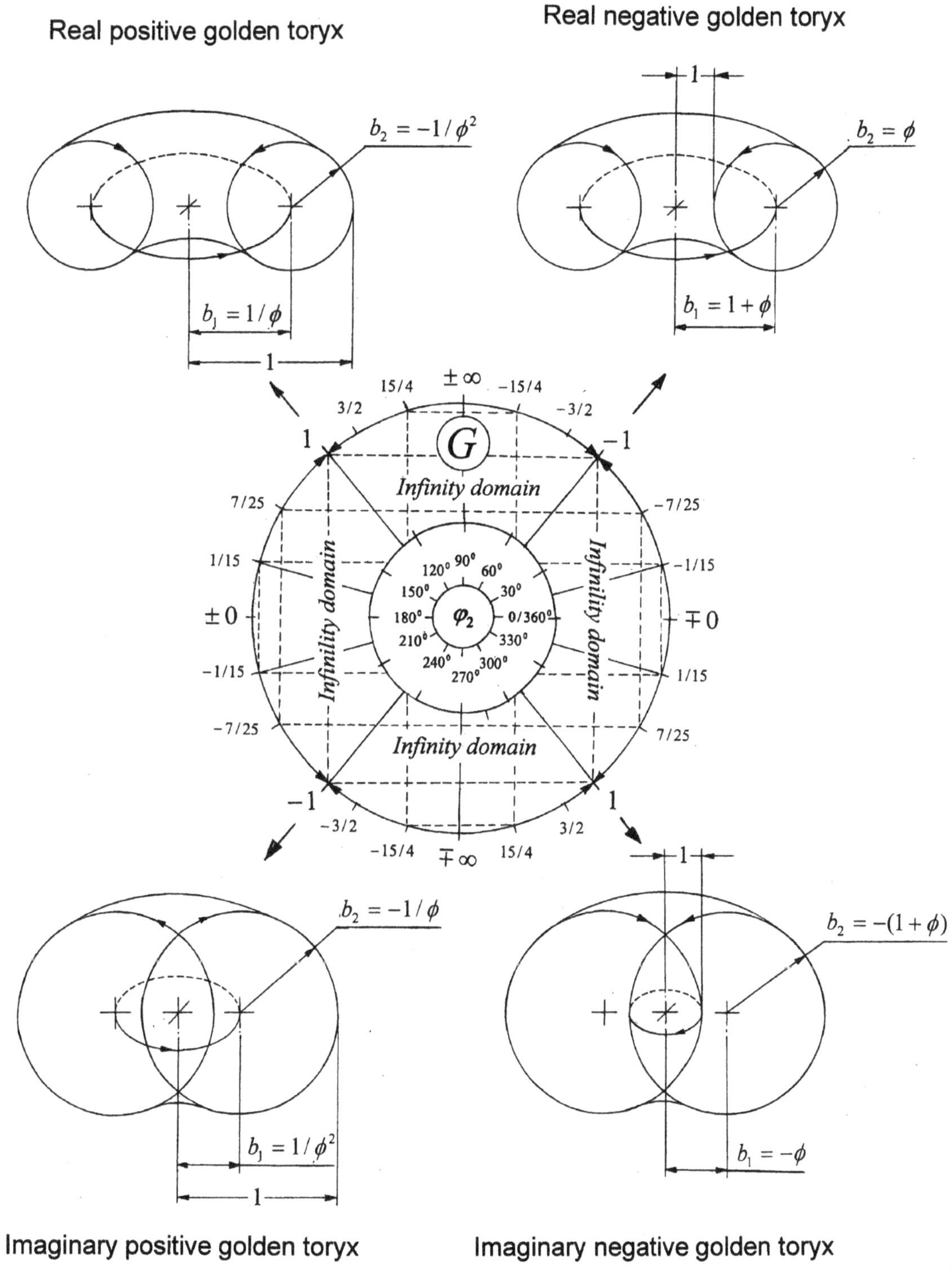
Real positive golden toryx
Real negative golden toryx
$b_2 = -1/\phi^2$
$b_1 = 1/\phi$
1
$b_2 = \phi$
$b_1 = 1 + \phi$
G
Infinity domain
Infinity domain
Infinity domain
Infinity domain
φ_2
±∞
∓∞
±0
∓0
Imaginary positive golden toryx
Imaginary negative golden toryx
$b_2 = -1/\phi$
$b_1 = 1/\phi^2$
$b_2 = -(1+\phi)$
$b_1 = -\phi$

Figure 5.8 Golden toryces corresponding to the toryx golden polarization $G = \pm 1$.

<u>*Notes*</u>

PART 2

Applied
Mathematics
of a Toryx

6. QUANTUM STATES OF TORYCES

CONTENTS

6.1 Definition of Quantum States of Excited Toryces

Toryces change their dimensions in quantum steps by a so-called *excitation process*. As shown in Figure 6.1, during the excitation of a toryx the radius of toryx leading string r_1 increases, while its eye radius r_0 remains constant.

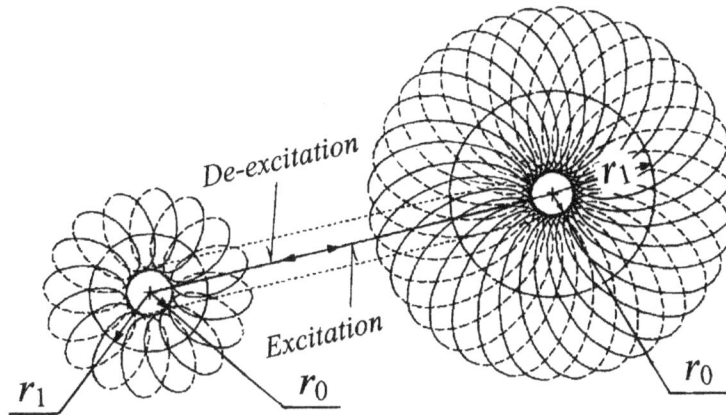

Figure 6.1. Excitation and de-excitation of toryces.

The derived quantization equations are based on three proposed limitations of degrees of freedom of real negative *lambda, harmonic* and *golden toryces* shown in Exhibit 6.1.

Exhibit 6.1. Limitations of degrees of freedom of excited toryces.

The relative radius of leading string of real negative toryx b_1 is equal to:		
Lambda toryx	**Harmonic toryx**	**Golden toryx**
$b_1 = z = 2(n\Lambda)^m$ (6.1-1)	$b_1 = z = 2 + n$ (6.1-2)	$b_1 = z = 2 + n/\phi$ (6.1-3)

In Exhibit 6.1:

z = toryx quantization parameter
$m \to 0, 1, 2, . .$, toryx exponential excitation quantum states
$n \to 0, 1, 2, . .$, toryx linear excitation quantum states
$\Lambda = 137, ...$, *toryx quantization constant*.

Notably, when $m \to 0$ and $n \to 0$, all three Eqs. (6.1-1) - (6.1-3) yield $z = 2$.

6.2 Quantization Equations for Excited Toryces

Table 6.2 shows quantization equations for relative radii of leading strings and spherical boundaries of excited toryces.

Table 6.2. Quantization equations for relative radii of toryx leading strings and spherical boundaries of excited toryces.

Lambda toryx	Harmonic toryx	Golden toryx	Relative radius of leading string	Relative radius of spherical boundary
$E^-_{m,n,q}$	$E^-_{H,n,q}$	$E^-_{G,n,q}$	$b^-_{1E} = z$ (6.2-1)	$b^-_E = 2z - 1$ (6.2-2)
$\breve{E}^-_{m,n,q}$	$\breve{E}^-_{H,n,q}$	$\breve{E}^-_{G,n,q}$	$\breve{b}^-_{1E} = 1 - z$ (6.2-3)	$\breve{b}^-_E = 1 - 2z$ (6.2-4)
$E^+_{m,n.q}$	$E^+_{H,n.q}$	$E^+_{G,n,q}$	$b^+_{1E} = \dfrac{z}{2z-1}$ (6.2-5)	$b^+_E = \dfrac{1}{2z-1}$ (6.2-6)
$\breve{E}^+_{m,n.q}$	$\breve{E}^+_{H,n.q}$	$\breve{E}^+_{G,n,q}$	$\breve{b}^+_{1E} = \dfrac{1-z}{1-2z}$ (6.2-7)	$\breve{b}^+_E = \dfrac{1}{1-2z}$ (6.2-8)
$A^-_{m,n,q}$	$A^-_{H,n.q}$	$A^-_{G,n,q}$	$b^-_{1A} = \dfrac{z}{z-1}$ (6.2-9)	$b^-_A = \dfrac{z+1}{z-1}$ (6.2-10)
$A^+_{m,n.q}$	$A^+_{H,n.q}$	$A^+_{G,n,q}$	$b^+_{1A} = \dfrac{z}{z+1}$ (6.2-11)	$b^+_A = \dfrac{z-1}{z+1}$ (6.2-12)
$\breve{A}^-_{m,n,q}$	$\breve{A}^-_{H,n.q}$	$\breve{A}^-_{G,n,q}$	$\breve{b}^-_{1A} = \dfrac{1}{1-z}$ (6.2-13)	$\breve{b}^-_A = \dfrac{1+z}{1-z}$ (6.2-14)
$\breve{A}^+_{m,n.q}$	$\breve{A}^+_{H,n.q}$	$\breve{A}^+_{G,n,q}$	$\breve{b}^+_{1A} = \dfrac{1}{1+z}$ (6.2-15)	$\breve{b}^+_A = \dfrac{1-z}{1+z}$ (6.2-16)

The first subscripts in the symbols of toryces are:
    For lambda toryces: m is standing for the toryx exponential excitation quantum state
    For harmonic toryces: H is standing for harmonic toryces
    For golden toryces: G is standing for golden toryces.
The second subscript indicates the toryx linear excitation quantum state n.
The third subscript indicates the oscillation quantum states q to be described in Section 6.4.
The superscript indicates the sign of toryx vorticities V and, in some particular cases, their values.

6.3 Quantization Equations for Resonant Excited Lambda Toryces

In the resonant toryx $\breve{E}r^-_{m,n,q}$, the frequency of its trailing string has the same amplitudes but opposite signs in respect to the basic toryx $E^-_{m,n,q}$. Their quantization equations are shown in Table 6.3.

Table 6.3. Quantization equations for the relative radii of leading strings and spherical boundary of basic and resonant lambda excited toryces.

Toryces	Relative radius of leading string	Relative radius of spherical boundary
Basic toryx $E^-_{m,n,q}$	$b^-_{1E} = z$ (6.2-1)	$b^-_E = 2z - 1$ (6.2-2)
Resonant toryx $\breve{E}r^-_{m,n,q}$	$\breve{b}^-_{1Er} = -z$ (6.3-1)	$\breve{b}^-_{Er} = -(2z - 1)$ (6.3-2)

6.4 Quantization Equations for Oscillated Toryces

During oscillation of a toryx, its radius of leading string r_1 and its eye radius r_0 change proportionally as shown in Fig. 6.4.1 in comparison with the toryx excitation during which the toryx eye radius r_0 remains constant, while the radius of its leading string r_1 increases.

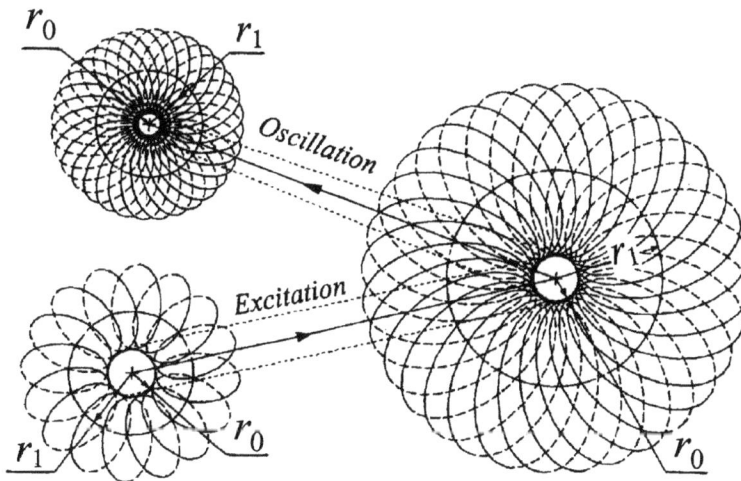

Figure 6.4.1. Oscillation of a toryx versus excitation of a toryx.

The oscillation of a toryx is a function of the ***toryx oscillation factor*** Q_q calculated based on the proposed limitation of its degree of freedom shown in Exhibit 6.4.

Exhibit 6.4. Limitations of degrees of freedom of oscillated toryces.

The toryx oscillation factor Q_q is equal to:

$$Q_0 = 1; \quad Q_q = 3\left(\frac{\Lambda}{2(q-1)}\right)^{q-1} \qquad (6.4\text{-}1)$$

where
$$q = 0, 1, 2, \ldots, \textit{toryx oscillation quantum states}.$$

The values of Q_q calculated from Eq. (6.4-1) are shown in Figure 6.4.2 and Table 6.4. At $q = q_m$ and $\Lambda = 137$, the toryx oscillation factor Q_q reaches its maximum value Q_{qm} defined by the equations:

$$q_m = 1 + \frac{\Lambda}{2e} = 26.199742$$

$$Q_{qm} = 3e^{\Lambda/2e} = 2.637728 \times 10^{11} \qquad (6.4\text{-}2)$$

When $q > q_m$, the toryx oscillation factor Q_q sharply decreases and at $q = 70.589986$ its magnitude reduces to 1. After that as q continues to increase and approaches infinity, Q_q decreases and approaches infinility.

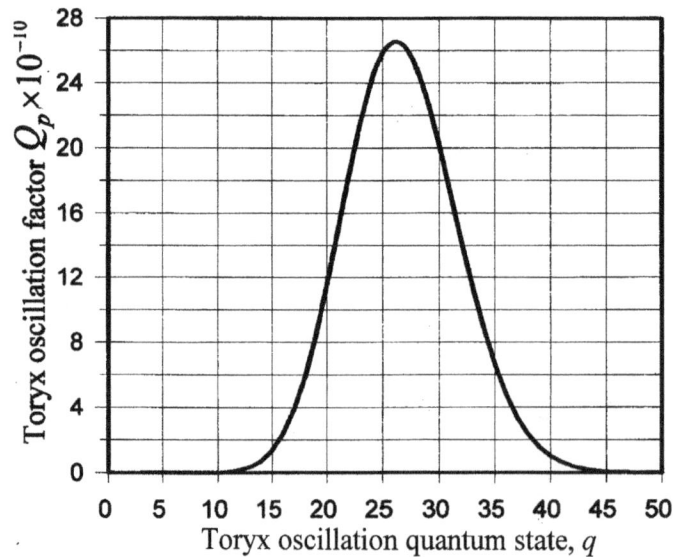

Figure 6.4.2. Toryx oscillation factor Q_q as a function of the toryx oscillation state q.

Table 6.4. Toryx oscillation factor Q_q as a function of the toryx oscillation quantum state q.

q	0	1	2	3	4	5	6
Q_q	1.000	3.000	205.500	3511.1875	35713.236	258014.18	1447851.734

6.5 Toryx Fine Structure

As shown in Figs. 6.5.1 and 6.5.2, fine structures of negative real and imaginary toryces are formed by *standing waves* oscillating inside *toryx fine-structure spherical boundary* with the radius r_3. Each standing wave is made up of a leading string with the radius $r_4 \to \infty$ and a double-helical trailing with the radius r_5.

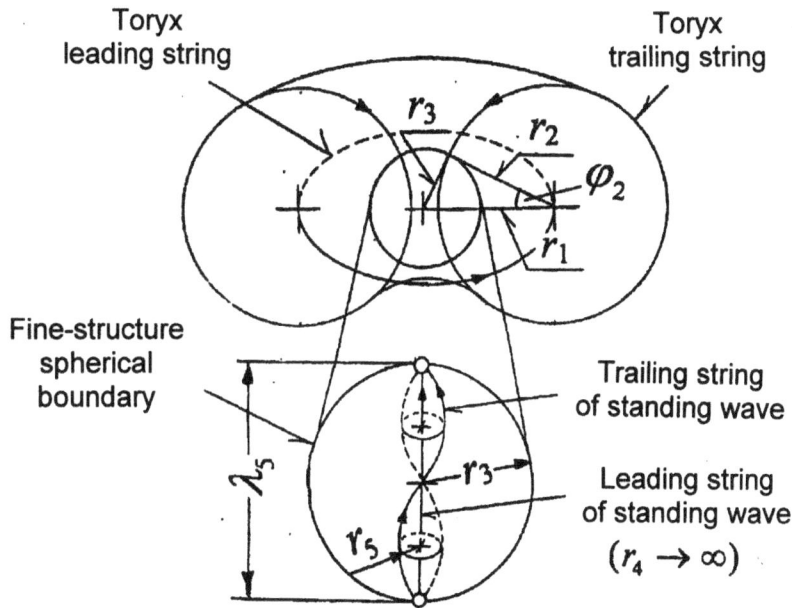

Figure 6.5.1. Fine structure of a real negative toryx.

The radius of toryx fine-structure spherical boundary r_3 is equal to:

$$r_3 = \sqrt{r_1^2 - r_2^2}$$

(6.5-1)

The relative radius of toryx fine-structure spherical boundary b_3 is equal to:

$$b_3 = \eta_2 = \sqrt{2b_1 - 1}$$

(6.5-2)

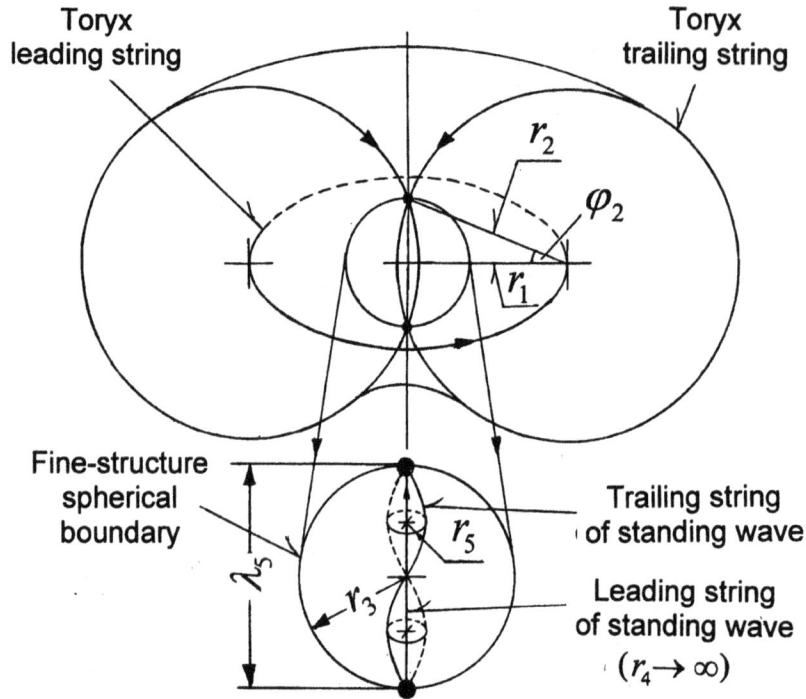

Figure 6.5.2. Fine structure of an imaginary negative toryx.

The perimeter $2\pi r_3$ of toryx fine-structure spherical boundary is equal to the wavelength λ_2 of toryx trailing string:

$$2\pi r_3 = \lambda_2 \tag{6.5-3}$$

The radius of leading string of toryx fine-structure standing wave r_4 approaches infinity:

$$r_4 \to \infty \tag{6.5-4}$$

The wavelengths of leading and trailing strings of toryx fine-structure standing wave, λ_4 and λ_5, are the same and equal to:

$$\lambda_4 = \lambda_5 = 2r_3 \tag{6.5-5}$$

The spiral length of trailing string of toryx fine-structure standing wave L_5 is related to its radius r_5 and wavelength λ_5 by the equation:

$$L_5 = \sqrt{\lambda_5^2 + (2\pi r_5)^2} \tag{6.5-6}$$

To define the radius r_5 of trailing string of toryx fine-structure standing wave, it is necessary to introduce the limitation of its degree of freedom shown in Exhibit 6.5.

Exhibit 6.5. Limitation of a degree of freedom of toryx fine-structure standing wave.

The aspect ratio of fine-structure standing wave $\lambda_5/2r_5$ is equal to the toryx quantization constant Λ:

$$\frac{\lambda_5}{2r_5} = \Lambda = const.$$

(6.5-7)

From Eqs. (6.5-5) and (6.5-6), the spiral length of trailing string of toryx fine-structure standing wave L_5 is equal to:

$$L_5 = \lambda_5 \frac{\alpha_s^{-1}}{\Lambda}$$

(6.5-8)

where α_s^{-1} is the **spacetime inverse fine-structure constant** related to the spacetime quantization constant Λ by the equation:

$$\alpha_S^{-1} = \sqrt{\Lambda^2 + \pi^2}$$

(6.5-9)

Notably, Eq. (6.5-9) is similar to the equation proposed by T.J. Burger as an approximation of the inverse fine structure constant. For the case when $\Lambda = 137$, this equation yields the value of $\alpha_S^{-1} = 137.036015720$. The relative difference between this values and the experimental value of the inverse fine structure constant $\alpha^{-1} = 137.035999074$ provided by 2011 CODATA is about one part per ten million.

Fine structure of toryx strings – The toryx strings are not absolute lines having no thicknesses, but the helical spirals with the radius r_s corresponding to the limits of spacetime defined by the equation:

$$r_s = \frac{l_p}{2\pi}$$

(6.5-10)

Where l_p is the Planck length.

<u>**Notes**</u>

7. FORMATION OF ELEMENTARY MATTER PARTICLES

7.1 Relationships between Physical & Spacetime Properties of Toryces

The following physical constants are used for describing toryx physical parameters:

c = velocity of light

e = elementary charge

l_p = Planck length

m_e = electron mass

m_p = proton mass

r_e = classical electron radius

μ_B = Bohr magneton

μ_N = nuclear magneton

α_s = spacetime fine structure constant

ε_0 = electric constant.

Derivation of relationships between toryx physical and spacetime parameters are based on two postulates shown in Exhibit 7.1.

Exhibit 7.1. Basic relationships between toryx physical and spacetime parameters.

- The relative toryx charge e_t / e_0 is equal to the toryx vorticity V:

$$\frac{e_t}{e} = V \qquad (7.1\text{-}1)$$

- The relative toryx gravitational mass m_g / m_e is proportional to the absolute value of the toryx vorticity $|V|$:

$$\frac{m_g}{m_e} = Q_q |V| \qquad (7.1\text{-}2)$$

Tables 7.1.1 and 7.1.2 show derived relationships between toryx physical and spacetime parameters.

Table 7.1.1. Relationships between toryx physical and spacetime parameters.

Toryx relative physical parameters	Equations	
Charge	$$\frac{e_t}{e} = V = -\frac{r_2}{r_1} = -\frac{b_1 - 1}{b_1} = \beta_{2r}$$	(7.1-3)
Gravitational mass	$$\frac{m_g}{Q_q m_e} = |V| = \left|\frac{r_2}{r_1}\right| = \left|\frac{b_1 - 1}{b_1}\right| = |\beta_{2r}|$$	(7.1-4)
Inertial mass	$$\frac{m_{ti}}{Q_q m_e} = \frac{2r_2}{r} = \frac{2b_2}{b} = \frac{2(b_1 - 1)}{2b_1 - 1} = \frac{2\delta_2 \beta_{2r}}{\beta_{2t}^2}$$	(7.1-5)
Magnetic moment	Real toryces: $$Q_q \frac{\mu_t}{\mu_B} = \pm\alpha \frac{VR}{2} = \pm\alpha \frac{(b_1 - 1)\sqrt{2b_1 - 1}}{2b_1}$$ Imaginary toryces: $$Q_q \frac{\breve{\mu}_t}{\mu_B} = \pm\alpha i \frac{VR}{2} = \pm\alpha i \frac{(\breve{b}_1 - 1)\sqrt{2\breve{b}_1 - 1}}{2\breve{b}_1}$$	(7.1-6)
Matter energy	$$\frac{E_m}{Q_q m_e c^2} = -V = \frac{b_1 - 1}{b_1}$$	(7.1-7)
Field energy	$$\frac{E_f}{Q_q m_e c^2} = \frac{1}{b_1} = \delta_2$$	(7.1-8)
Kinetic energy	$$\frac{K_t}{Q_q m_e c^2} = \frac{(b_1 - 1)}{b_1^2} = -V\delta_2$$	(7.1-9)
Potential energy	$$\frac{U_t}{Q_q m_e c^2} = -\frac{2(b_1 - 1)}{b_1^2} = 2V\delta_2$$	(7.1-10)
Total kinetic & potential energy	$$\frac{E_t}{Q_q m_e c^2} = -\frac{(b_1 - 1)}{b_1^2} = V\delta_2$$	(7.1-11)
Density	$$\rho_{tr} = \rho_t \frac{2\pi^2 r_0^3}{m_e Q_q} = \left|\frac{b_1 - 1}{b_1}\right| \frac{1}{b_1(b_1 - 1)^2}$$	(7.1-12)
Modulus of elasticity	$$B_{tr} = B_t \frac{2\pi^2 r_0^3}{Q_q m_e c^2} = \left|\frac{b_1 - 1}{b_1}\right| \frac{2b_1 - 1}{b_1^3 (b_1 - 1)^2}$$	(7.1-13)

Table 7.1.2. Relationships between toryx spacetime and physical parameters.

Toryx spacetime parameters	Equations	
Eye radius	$r_0 = \dfrac{r_e}{2} = \dfrac{e^2}{8\pi\varepsilon_0 m_e c^2}$	(7.1-14)
Base frequency	$f_0 = \dfrac{c}{2\pi r_0} = \dfrac{4\varepsilon_0 m_e c^3}{e^2}$	(7.1-15)
Base period	$T_0 = \dfrac{2\pi r_0}{c} = \dfrac{e^2}{4\varepsilon_0 m_e c^3}$	(7.1-16)

Toryx relative magnetic moment – Figure 7.1 shows a plot of Eq. (7.1-4) for the toryx relative magnetic moment μ_t / μ_B expressed in respect to the Bohr magneton μ_B given by the equation:

$$\mu_B = \frac{e^3}{8\pi\alpha_s \varepsilon_0 m_e c}$$
(7.1-15)

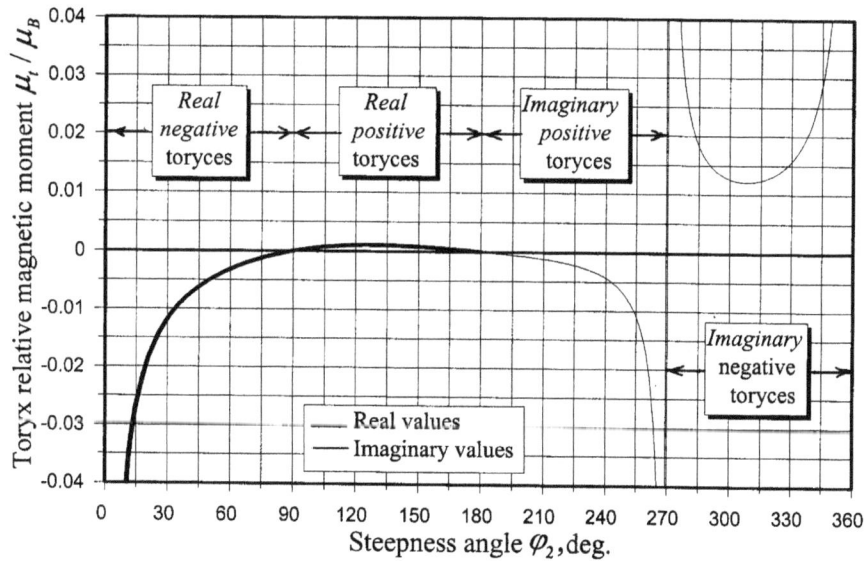

φ_2	$360^0/0^0$	90^0	180^0	270^0
μ_t / μ_B	$+\infty i/-\infty$	$-0/+0$	$+0/-0i$	$-\infty i/+\infty i$

Figure 7.1.1. Toryx relative magnetic moment in respect to the Bohr magneton μ_t / μ_B as a function of the steepness angle of trailing string φ_2 when $Q_q = 1$.

The toryx relative magnetic moment μ_t / μ_B is expressed in respect to the Bohr magneton μ_B that is given by the equation:

$$\mu_B = \frac{e^3}{8\pi\alpha_s\varepsilon_0 m_e c} \qquad (7.1\text{-}16)$$

The toryx relative magnetic moment in respect to the nuclear magneton μ_t / μ_N is related to the toryx relative magnetic moment in respect to the Bohr magneton μ_t / μ_B by the equation:

$$\frac{\mu_t}{\mu_N} = \frac{\mu_t}{\mu_B}\frac{m_p}{m_e} \qquad (7.1\text{-}17)$$

Toryx field & matter energy - Figure 7.1.2 shows plots of Eqs. (7.1-5) and (7.1-6) for the toryx relative field energy $E_f / Q_q m_e c^2$ and matter energy $E_m / Q_q m_e c^2$ as functions of the steepness angle of trailing string φ_2 when $Q_q = 1$.

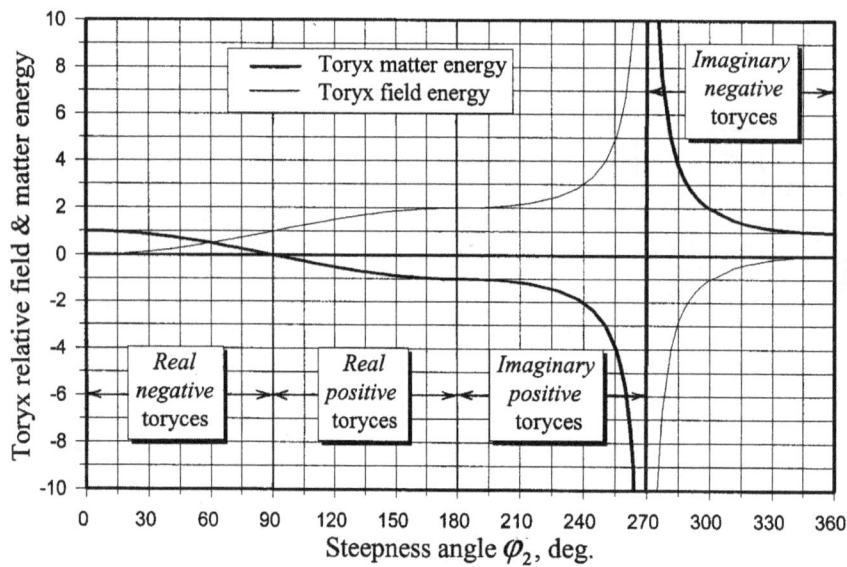

φ_2	$360^0/0^0$	90^0	180^0	270^0
$E_f / Q_q m_e c^2$	$-0/+0$	$+1$	$+2$	$+\infty/-\infty$
$E_m / Q_q m_e c^2$	$+1.0$	$+0/-0$	-1.0	$-\infty/+\infty$

Figure 7.1.2. Toryx relative field energy $E_f / Q_q m_e c^2$ and matter energy $E_m / Q_q m_e c^2$ as functions of the steepness angle of trailing string φ_2 when $Q_q = 1$.

Toryx density – Figure 7.1.3 shows a plot of Eq. (7.1-10) for the toryx relative density ρ_{tr}. It is derived based on the assumption equal to the ratio of the toryx gravitational mass m_{tg} to the volume occupied by its trailing string as given by the equation:

$$\rho_t = \frac{m_e Q_q}{2\pi^2 r_0^3} \left| \frac{b_1 - 1}{b_1} \right| \frac{1}{b_1(b_1 - 1)^2} \qquad (7.1\text{-}18)$$

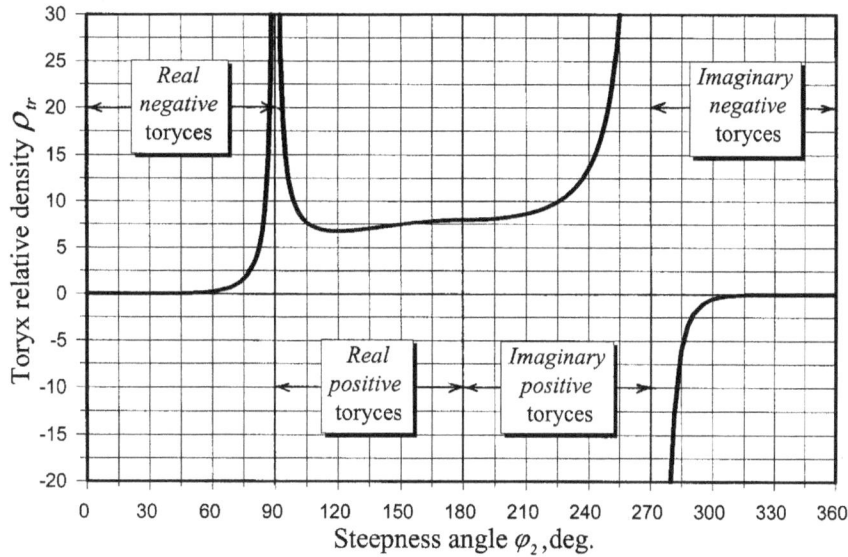

φ_2	$360^0 / 0^0$	90^0	180^0	270^0
ρ_{tr}	$-0/+0$	$+\infty$	8.0	$+\infty/-\infty$

Figure 7.1.3. Toryx relative density ρ_{tr} as a function of the steepness angle of trailing string φ_2 when $Q_q = 1$.

Toryx modulus of elasticity – Figure 7.1.4 shows a plot of Eq. (7.1-10) for the toryx relative modulus of elasticity B_{tr}. In classical mechanics, elastic properties, or rigidity, of compressible media contained in a certain volume are defined by the **bulk modulus of elasticity**. It is equal to the ratio of change in pressure inside the media to the resulting fractional change in the volume of the media. Let compression waves, like sound waves, travel through the toryx trailing string with velocity equal to the spiral velocity of the toryx leading string V_1. In that case, the toryx bulk modulus B_t will be equal to:

$$B_t = \rho_t V_1^2 = \frac{Q_q m_e c^2}{2\pi^2 r_0^3} \left| \frac{b_1 - 1}{b_1} \right| \frac{2b_1 - 1}{b_1^3 (b_1 - 1)^2} \qquad (7.1\text{-}19)$$

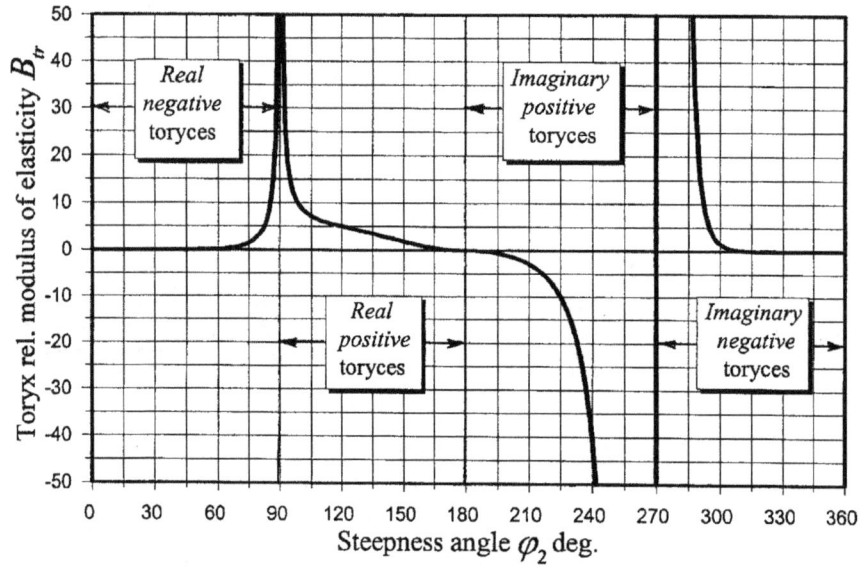

φ_2	$360^0 / 0^0$	90^0	180^0	270^0
B_{tr}	0	$+\infty$	$+0/-0$	$-\infty/+\infty$

Figure 7.1.4 Toryx relative bulk modulus of elasticity B_{tr} as a function of the steepness angle of trailing string φ_2.

Toryx relativistic equations are shown in Table 7.1.4.

- Both masses and charges of toryces are dependent on velocities of their leading strings.
- In real toryces the absolute values of their relative charges and masses decrease with an increase of the relative spiral velocities of their leading strings β_1.
- In imaginary toryces the absolute values of their relative charges and masses increase as the relative spiral velocities of their leading string β_1 increase.

Table 7.1.4. Toryx relativistic equations.

Relative charge	$\dfrac{e_t}{e} = \sqrt{1 - \beta_1^2}$	(7.1-20)
Relative gravitational mass	$\dfrac{m_{tg}}{m_e} = Q_q \left\| \sqrt{1 - \beta_1^2} \right\|$	(7.1-21)

7.2 Formation of Virtual Toryces

According to the UST, the toryx spacetime limits correspond to the Planck length l_p at which the toryx spacetime properties are no longer governed by the toryx spacetime postulates, but by the rules of uncertainty of **quantum vacuum**. Consequently, the primordial toryces are produced spontaneously in quantum vacuum. The process is governed by the proposed **toryx uncertainty principle** presented in Exhibit 7.2.

Exhibit 7.2. Toryx uncertainty principle.

A virtual toryx can be produced spontaneously in quantum vacuum and exist for the period equal to the relative period of its leading string t_1 if an absolute product of the toryx vorticity V and the period t_1 is equal or less than $1/(Q_q \pi \alpha_s)$ as given by the equation:

$$\text{Real toryces: } |Vt_1| = \frac{b_1(b_1-1)}{\sqrt{2b_1-1}} \leq \frac{1}{Q_q \pi \alpha_s} \qquad (7.2\text{-}1)$$

$$\text{Imaginary toryces: } |Vt_1| = \frac{b_1(b_1-1)i}{\sqrt{2b_1-1}} \leq \frac{1}{Q_q \pi \alpha_s} \qquad (7.2\text{-}2)$$

For the non-oscillated harmonic toryces $(Q_q = 1)$, we obtain from Eqs. (7.2-1) and (7.2-2):

$$0.5000164241 < b_1 < 16.1229570320 \qquad (7.2\text{-}3)$$

$$0.4999835759 > b_1 > -15.1229570320 \qquad (7.2\text{-}4)$$

We called the toryces formed in quantum vacuum the **virtual toryces**. Figure 7.2 shows ranges of the toryx relative radii of spherical boundaries b within which the spacetime ceases to exist and replaced by spontaneously appearing virtual toryces of quantum vacuum.

The relative spacetime length b_s corresponding to the limits of spacetime is equal to:

$$b_s = \frac{l_p}{2\pi r_0} \qquad (7.2\text{-}5)$$

Table 7.2 shows the ranges of the toryx relative radii of spherical boundaries b, leading string b_1 and trailing string b_2 of electrons, positrons, ethertrons and singulatrons within which the spacetime ceases to exist and replaced by the virtual particles of quantum vacuum.

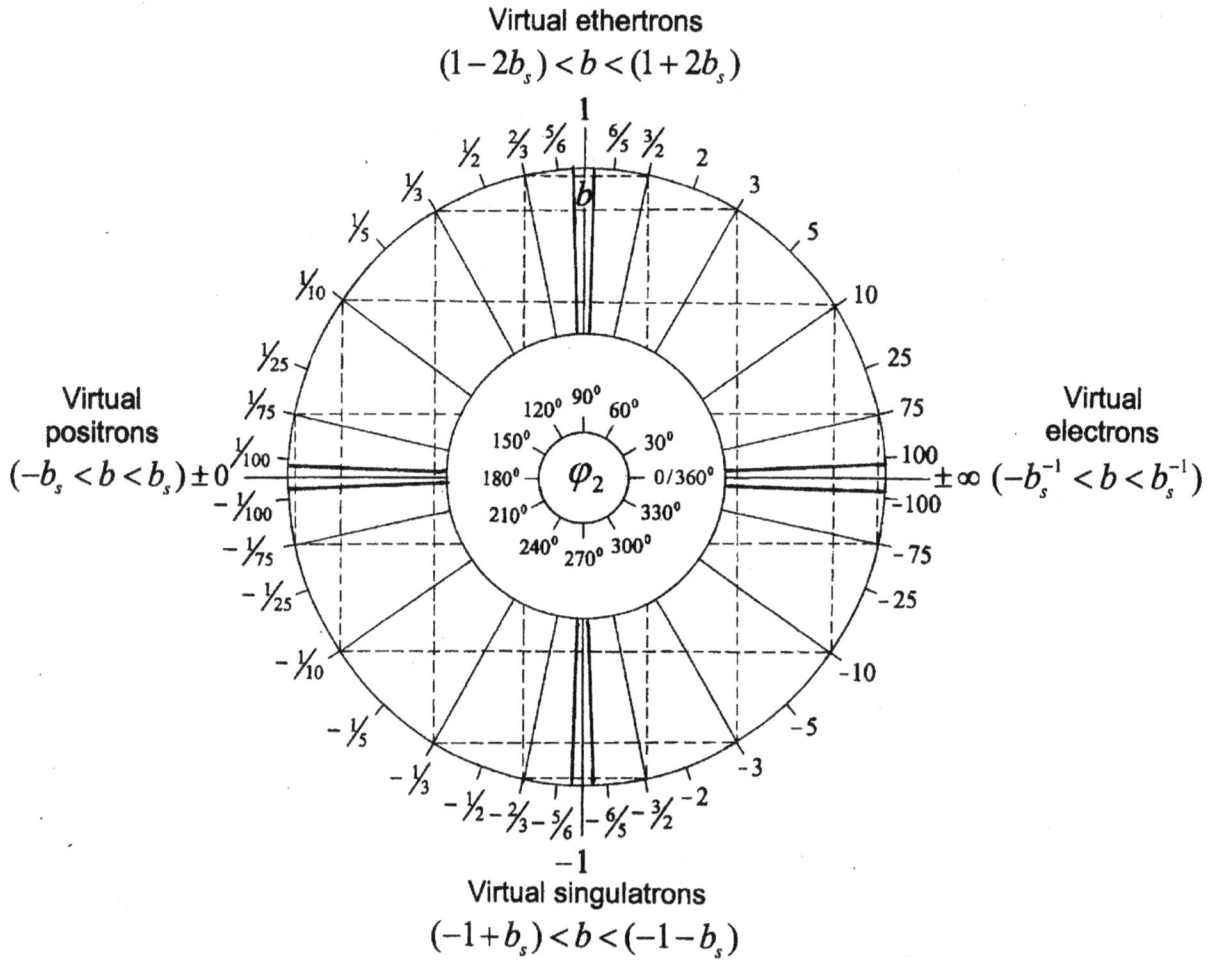

Figure 7.2. Ranges of the toryx relative radii of spherical boundaries b within which the spacetime ceases to exist and replaced by the virtual particles of quantum vacuum.

Table 7.2. Ranges of toryx parameters within which spacetime ceases to exist and replaced by quantum vacuum.

Virtual particles	Leading string	Trailing string	Spherical boundary
Electron	$\dfrac{1-b_s^{-1}}{2} < b_1 < \dfrac{1+b_s^{-1}}{2}$	$-\dfrac{1+b_s^{-1}}{2} < b_2 < -\dfrac{1-b_s^{-1}}{2}$	$-b_s^{-1} < b < b_s^{-1}$
Positron	$\dfrac{1-b_s}{2} < b_1 < \dfrac{1+b_s}{2}$	$-\dfrac{1+b_s}{2} < b_2 < -\dfrac{1-b_s}{2}$	$-b_s < b < b_s$
Ethertron	$1-b_s < b_1 < 1+b_s$	$-b_s < b_2 < b_s$	$1-2b_s < b < 1+2b_s$
Singulatron	$-b_s < b_1 < b_s$	$-1-b_s < b_2 < -1+b_s$	$-1-2b_s < b < -1+2b_s$

7.3   From Quantum Vacuum to Elementary Matter Particles

Four basic elementary particles (trons) are formed from quantum vacuum by the unification of adjacent toryces: *electrons, positrons, ethertrons* and *singulatrons* as shown in Fig. 7.3.1.

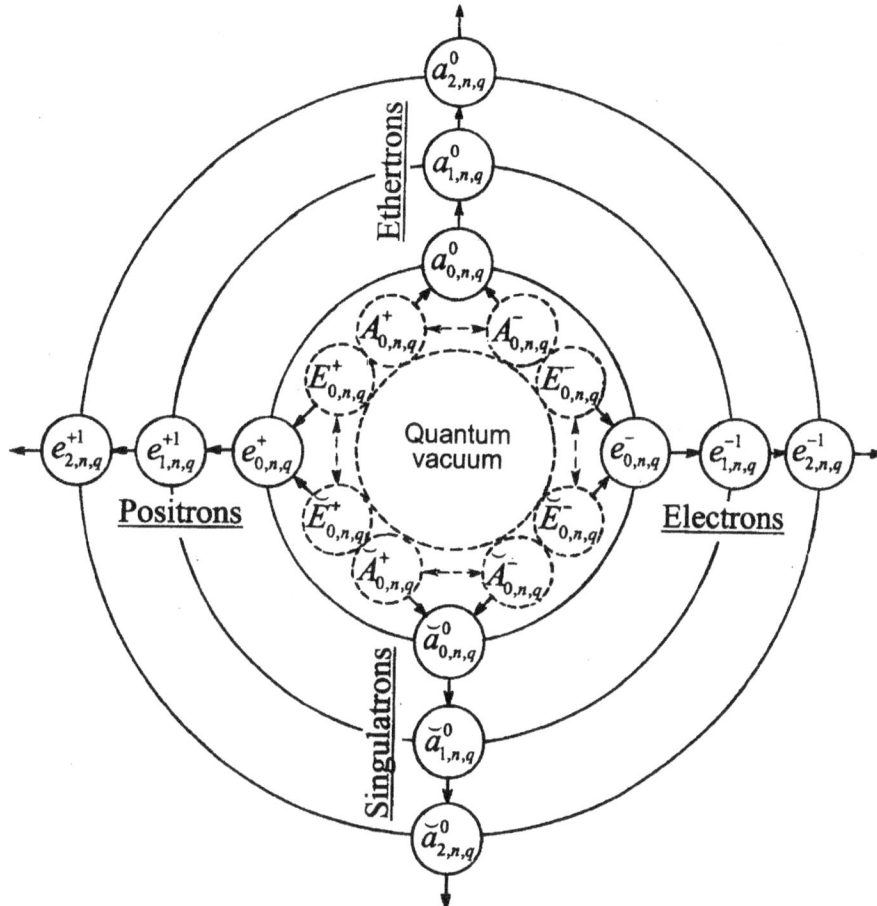

Figure 7.3.1. Formation of elementary particles (trons) from quantum vacuum.

The elementary particles (trons) are created in four steps:

1. Formation of *virtual toryces* from *quantum vacuum*
2. Formation of *harmonic toryces* from virtual toryces
3. Formation of *harmonic elementary particles* by unification
    of polarized harmonic toryces
4. Formation of *excited and oscillated elementary particles*
    from harmonic toryces.

The trons are divided into two kinds: *self-polarized trons and mutually-polarized trons*.

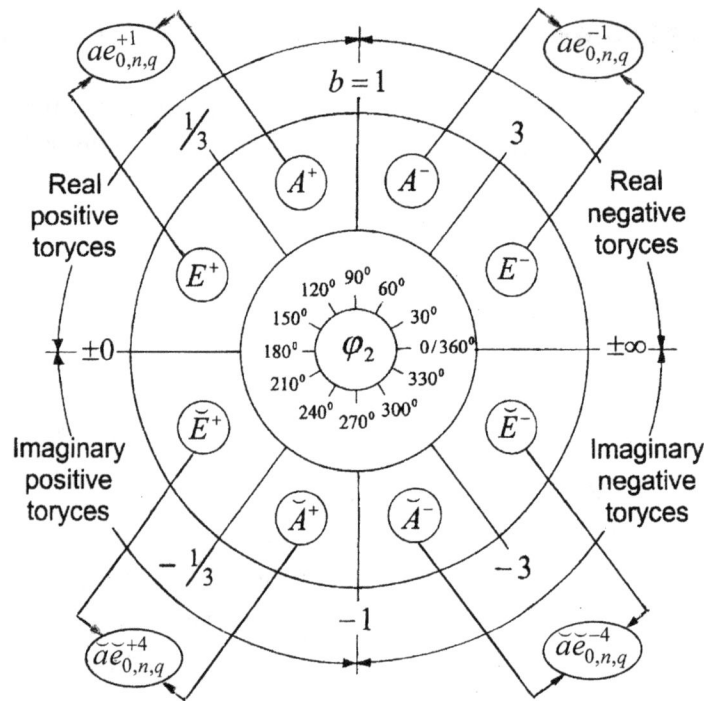

Figure 7.3.2. Formation of self-polarized harmonic lambda trons.

Table 7.3.1. Constituent toryces of self-polarized harmonic lambda trons.

Trons	Constituent toryces	Trons	Constituent toryces
$ae_{0,n,q}^{-1}$	$A_{0,n,q}^{-1/2} + E_{0,n,q}^{-1/2}$	$ae_{0,n,q}^{+1}$	$A_{0,n,q}^{+1/2} + E_{0,n,q}^{+1/2}$
$\breve{\breve{ae}}_{0,n,q}^{-4}$	$\breve{A}_{0,n,q}^{-2} + \breve{E}_{0,n,q}^{-2}$	$\breve{\breve{ae}}_{0,n,q}^{+4}$	$\breve{A}_{0,n,q}^{+2} + \breve{E}_{0,n,q}^{+2}$

Figure 7.3.2 and Table 7.3.1 show general presentations of formation of self-polarized harmonic lambda trons.

- The self-polarized real lambda electrons $ae_{0,n,q}^{-1}$ are made up of the real negative harmonic lambda toryces $A_{0,n,q}^{-1/2}$ and $E_{0,n,q}^{-1/2}$.

- The self-polarized real lambda positrons $ae_{0,n,q}^{+1}$ are made up of the real positive harmonic lambda toryces $A_{0,n,q}^{+1/2}$ and $E_{0,n,q}^{+1/2}$.

- The self-polarized imaginary lambda trons $\breve{\breve{ae}}_{0,n,q}^{-4}$ are made up of the imaginary negative harmonic lambda toryces $\breve{A}_{0,n,q}^{-2}$ and $\breve{E}_{0,n,q}^{-2}$.

- The self-polarized imaginary lambda trons $ae_{0,n,q}^{+4}$ are made up of the imaginary positive harmonic lambda toryces $A_{0,n,q}^{+2}$ and $E_{0,n,q}^{+2}$.

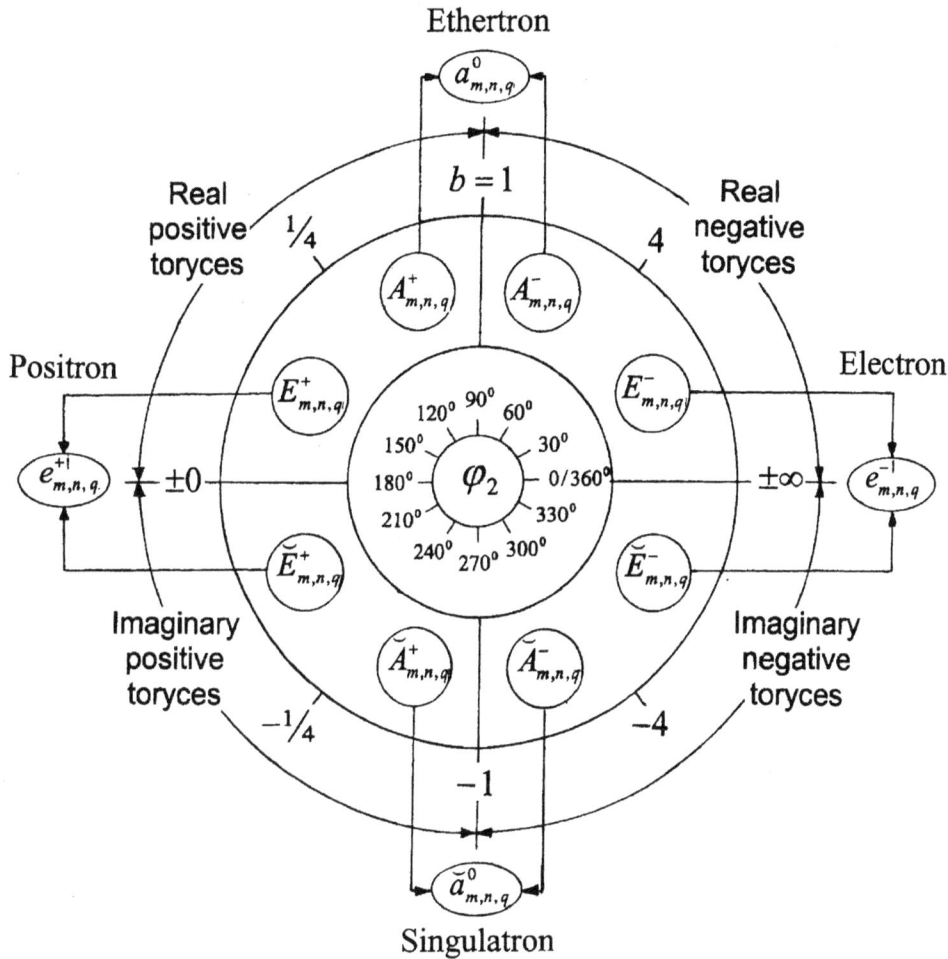

Figure 7.3.3. Formation of mutually-polarized excited lambda trons.

Table 7.3.2. Constituent toryces of mutually-polarized excited lambda trons.

Tron	Constituent toryces	Tron	Constituent toryces
Electron $e_{m,n,q}^{-1}$	$E_{m,n,q}^{-} + \breve{E}_{m,n,q}^{-}$	**Ethertron** $a_{m,n,q}^{0}$	$A_{m,n,q}^{-} + A_{m,n,q}^{+}$
Positron $e_{m,n,q}^{+1}$	$E_{m,n,q}^{+} + \breve{E}_{m,n,q}^{+}$	**Singulatron** $\breve{a}_{m,n,q}^{0}$	$\breve{A}_{m,n,q}^{-} + \breve{A}_{m,n,q}^{+}$

Figure 7.3.3 and Table 7.3.2 show general presentations of formation of four mutually-polarized trons.

- The electrons $e^{-1}_{m,n,q}$ are reality-polarized negative trons; they are made up of the real negative toryces $E^-_{m,n,q}$ and the imaginary negative toryces $\breve{E}^-_{m,n,q}$.

- The positrons $e^{+1}_{m,n,q}$ are reality-polarized positive trons; they are made up of the real positive toryces $E^+_{m,n,q}$ and the imaginary positive toryces $\breve{E}^+_{m,n,q}$.

- The ethertrons $a^0_{m,n,q}$ are vorticity-polarized real trons; they are made up of the real negative toryces $A^-_{m,n,q}$ and the real positive toryces $A^+_{m,n,q}$.

- The singulatrons $\breve{a}^0_{m,n,q}$ are vorticity-polarized imaginary trons; are made up of the imaginary negative toryces $\breve{A}^-_{m,n,q}$ and the imaginary positive toryces $\breve{A}^+_{m,n,q}$.

7.4 Rules of Formation of Stable Elementary Matter Particles

Trons sustain their existence by a periodic absorption and release of spacetime. Imaginary toryces are responsible for absorption of spacetime, while real toryces are responsible for release of spacetime. Stability of reality-polarized trons is governed by the proposed ***tron polarization conservation law*** (see Exhibit 7.4).

Exhibit 7.4. Tron polarization conservation law.

The ***toryx integral polarization*** I is given by the equation:

$$I = Q_q V R^2 = -\frac{Q_q(b_1 - 1)(2b_1 - 1)}{b_1} \tag{7.4-1}$$

The ***integral polarization*** I_t of a stable tron made up of N real toryces with the integral polarization I and $\breve{N}$ imaginary toryces with the integral polarization $\breve{I}$ is infinitesimally small:

$$I_t = (IN + \breve{I}\breve{N}) \to \pm 0 \tag{7.4-2}$$

Consequently, in a stable tron made up of N real and $\breve{N}$ imaginary toryces, the ***stable tron reality ratio*** T is defined by the equation:

$$T = \frac{N}{\breve{N}} = \left(\frac{b_1}{b_1 - 1}\right)^2 = \left(\frac{b+1}{b-1}\right)^2 \tag{7.4-3}$$

where b_1 and b are respectively relative radii of toryx leading string and spherical boundary.

The following additional rules are applied to the toryces forming mutually-polarized and self-polarized trons:

- Mutually-polarized toryces coexist **consecutively** inside their trons, so the tron parameters are equal to the arithmetic average sums of respective parameters of their constituent toryces as expressed below by Eq. (7.4-4) in application to the tron vorticity V_t and in Eq. (7.4-5) in application to the tron reality R_t:

$$V_t = 0.5(V^+ N^+ + V^- N^-) \tag{7.4-4}$$

$$R_t = 0.5(R\,N + \breve{R}\,\breve{N}i) \tag{7.4-5}$$

- The effective rotational velocities $\breve{V}_{1r} = \breve{V}_{2t}$ of toryx leading and trailing strings of imaginary toryces are equal to:

$$\breve{V}_{1r} = \breve{V}_{2t} = V_{1r}i = V_{2t}i \tag{7.4-6}$$

- Self-polarized toryces coexist **concurrently** inside their trons, so the tron parameters are equal to the arithmetic sums of vorticities and realities of their respective constituent toryces.
- The exponential excitation quantum states m of toryces forming trons depend on the spacetime levels L of atoms that are formed from trons as shown in Table 7.4, where s is the spacetime level number.

Table 7.4. Spacetime levels of atoms.

Spacetime levels of atoms	Tron quantum states m			
	Ethertron	**Singulatron**	**Electron**	**Positron**
L0 (harmonic matter)	$m = 0$	$m = 0$	$m = 0$	$m = 0$
L1	$m = 0$	$m = 0$	$m = 1$	$m = 1$
L2 (ordinary matter)	$m = 1$	$m = 1$	$m = 2$	$m = 2$
L3	$m = 2$	$m = 2$	$m = 3$	$m = 3$
. . . .	. . . .	. . . .	. . . .	. . . .

<u>*Notes*</u>

8. EXAMPLES OF ELEMENTARY MATTER PARTICLES

8.1 Reality-polarized harmonic lambda electron $e_{0,0,0}^{-1}$ is made up of two harmonic real negative toryces $E_{0,0,0}^{-\frac{1}{2}}$ and one half of the harmonic imaginary negative toryx $\breve{E}_{0,0,0}^{-2}$ as shown by the equation:

$$e_{0,0,0}^{-1} = 2E_{0,0,0}^{-\frac{1}{2}} + \tfrac{1}{2}\breve{E}_{0,0,0}^{-2}$$

Figure 8.1 and Table 8.1 show cross-section, dimensions, compositions and properties, of the reality-polarized harmonic lambda electron $e_{0,0,0}^{-1}$.

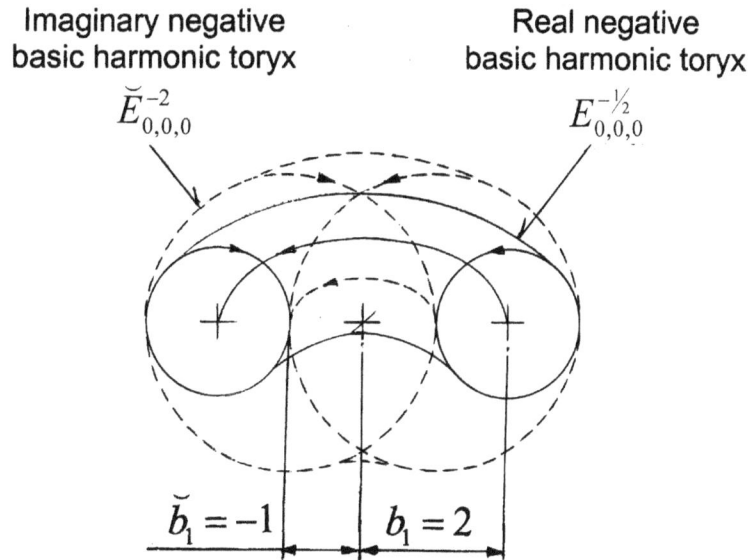

Figure 8.1. Cross-section and dimensions of the reality-polarized harmonic lambda electron $e_{0,0,0}^{-1}$ and its constituent toryces.

Table 8.1. Composition and properties of the reality-polarized harmonic lambda electron $e_{0,0,0}^{-1}$ and its constituent toryces.

Toryx	b_1	$\beta_1 = \beta_{2t}$	w_2	e_t / e	μ_t / μ_N	m_{tg} / m_e	I	N
$E_{0,0,0}^{-\frac{1}{2}}$	2.0	0.866025	1.154701	-0.50	- 5.80196032	0.500	+1.5	2
$\breve{E}_{0,0,0}^{-2}$	-1.0	-1.732051i	- 0.577350i	-2.00	- 23.20784127i	2.000	- 6	0.5
Harmonic lambda electron $e_{0,0,0}^{-1}$				**-1.00**	**-11.60392064**	**1.000**	**0.00**	**1**

8.2 *Reality-polarized resonant harmonic lambda electron* $er_{0,0,0}^{-1}$ is made up of one harmonic real negative toryces $E_{0,0,0}^{-1/2}$ and one resonant oscillated harmonic imaginary negative toryx $\breve{E}r_{0,0,0}^{-3/2}$ as shown by the equation:

$$er_{0,0,0}^{-1} = E_{0,0,0}^{-1/2} + \breve{E}r_{0,0,0}^{-3/2}$$

Figure 8.2 and Table 8.2 show cross-section, dimensions, compositions and properties of the reality-polarized resonant harmonic lambda electron $er_{0,0,0}^{-1}$.

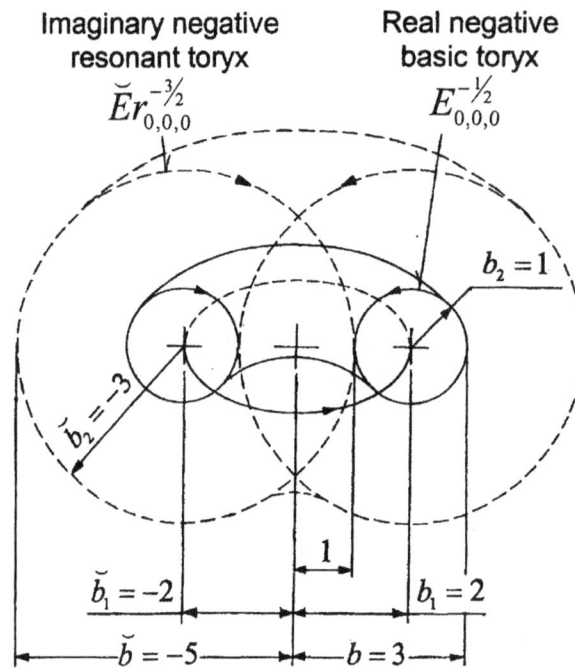

Figure 8.2. Cross-section and dimensions of the reality-polarized resonant harmonic lambda electron $er_{0,0,0}^{-1}$ and its constituent toryces.

Table 8.2. Composition and properties of the reality-polarized resonant harmonic lambda electron $er_{0,0,0}^{-1}$ and its constituent toryces.

Toryx	b_1	$\beta_1 = \beta_{2t}$	w_2	e_t/e	μ_t/μ_N	m_{tg}/m_e	I	N
$E_{0,0,0}^{-1/2}$	2.0	0.866025	1.154701	-0.50	- 5.80196032	0.500	+1.5	1
$\breve{E}r_{0,0,0}^{-3/2}$	-2.0	-1.118034i	- 0.894427	-1.50	- 22.47089569i	1.500	-7.5	1
Resonant harmonic lambda electron $er_{0,0,0}^{-1}$				**-1.00**	**-14.13642800**	**1.000**	**-3.00**	**1**

8.3 Reality-polarized harmonic lambda positron $e_{0,0,0}^{+1}$ is made up of two harmonic real positive toryces $E_{0,0,0}^{+\frac{1}{2}}$ and one half of the harmonic imaginary positive toryx $\breve{E}_{0,0,0}^{+2}$ as shown by the equation:

$$e_{0,0,0}^{+1} = 2E_{0,0,0}^{+\frac{1}{2}} + \frac{1}{2}\breve{E}_{0,0,0}^{+2}$$

Figure 8.3 and Table 8.3 show cross-section, dimensions, compositions and properties of the reality-polarized harmonic lambda positron $e_{0,0,0}^{+1}$.

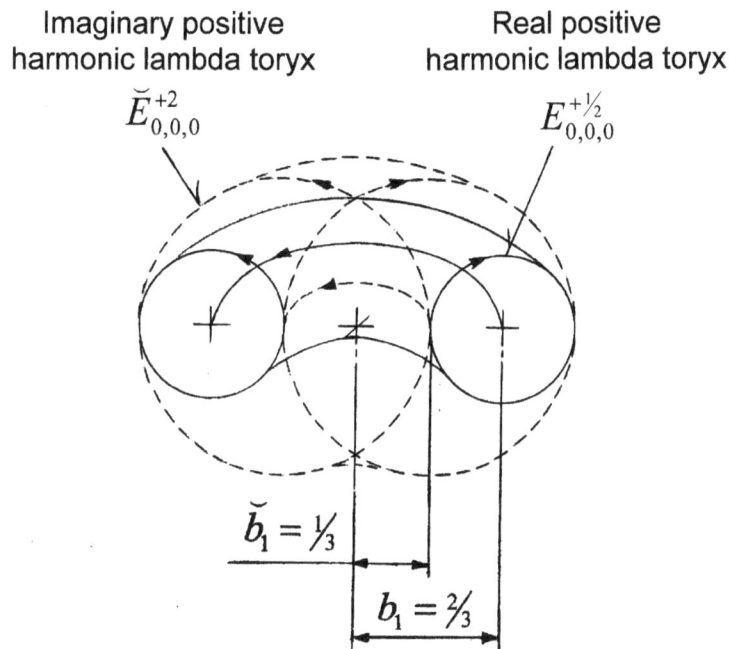

Figure 8.3. Cross-section and dimensions of the reality-polarized basic harmonic positron $e_{0,0,0}^{+1}$ and its constituent toryces.

Table 8.3. Composition and properties of the reality-polarized harmonic lambda positron $e_{0,0,0}^{+1}$ and its constituent toryces.

Toryx	b_1	$\beta_1 = \beta_{2t}$	w_2	e_t/e	μ_t/μ_N	m_{tg}/m_e	I	N
$E_{0,0,0}^{+\frac{1}{2}}$	$\frac{2}{3}$	0.866025	1.154701	+ 0.50	+1.93398677	0.500	$-\frac{1}{6}$	2
$\breve{E}_{0,0,0}^{+2}$	$\frac{1}{3}$	- 1.732051i	0.577350i	+ 2.00	+8.73594709i	2.000	$+\frac{2}{3}$	0.5
Harmonic lambda positron $e_{0,0,0}^{+1}$				**+1.00**	**+3.86797355**	**1.000**	**0.00**	**1**

8.4 Vorticity-polarized harmonic lambda ethertron $a_{0,0,0}^{0}$ is made up of one harmonic real negative toryx $A_{0,0,0}^{-\frac{1}{2}}$ and one harmonic real positive toryx $A_{0,0,0}^{+\frac{1}{2}}$ as shown by the equation:

$$a_{0,0,0}^{0} = A_{0,0,0}^{-\frac{1}{2}} + A_{0,0,0}^{+\frac{1}{2}}$$

Figure 8.4 and Table 8.4 show cross-section, dimensions, compositions and properties of the harmonic lambda ethertron $a_{0,0,0}^{0}$.

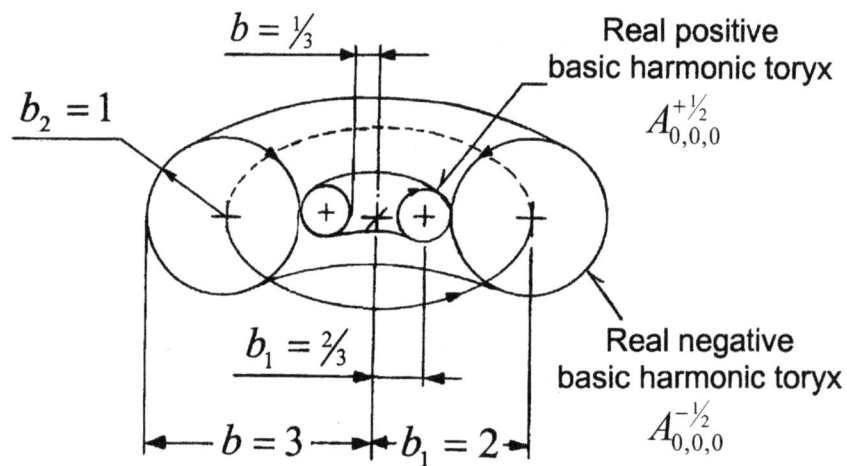

Figure 8.4. Cross-section and dimensions of the vorticity-polarized harmonic lambda ethertron $a_{0,0,0}^{0}$ and its constituent toryces.

Table 8.4. Composition and properties of the vorticity-polarized harmonic ethertron $a_{0,0,0}^{0}$ and its constituent toryces.

Toryx	b_1	$\beta_1 = \beta_{2t}$	w_2	e_t/e	μ_t/μ_N	m_{tg}/m_e	I	N
$A_{0,0,0}^{-\frac{1}{2}}$	2.0	0.866025	1.154701	-0.50	-5.80196032	0.500	+1.5	1
$A_{0,0,0}^{+\frac{1}{2}}$	$\frac{2}{3}$	0.866025	1.154701	+0.50	+1.93398677	0.500	$-\frac{1}{6}$	1
Harmonic lambda ethertron $a_{0,0,0}^{0}$				0.00	-1.93398677	0.500	$+\frac{2}{3}$	1

8.5 Vorticity-polarized harmonic lambda singulatron $\breve{a}^0_{0,0,0}$ is made up one harmonic imaginary negative toryx $\breve{A}^{-2}_{0,0,0}$ and one harmonic imaginary positive toryx $\breve{A}^{+2}_{0,0,0}$ as shown by the equation:

$$\breve{a}^0_{0,0,0} = \breve{A}^{-2}_{0,0,0} + \breve{A}^{+2}_{0,0,0}$$

Figure 8.5 and Table 8.5 show cross-section, dimensions, compositions and properties of the harmonic lambda singulatron $\breve{a}^0_{0,0,0}$.

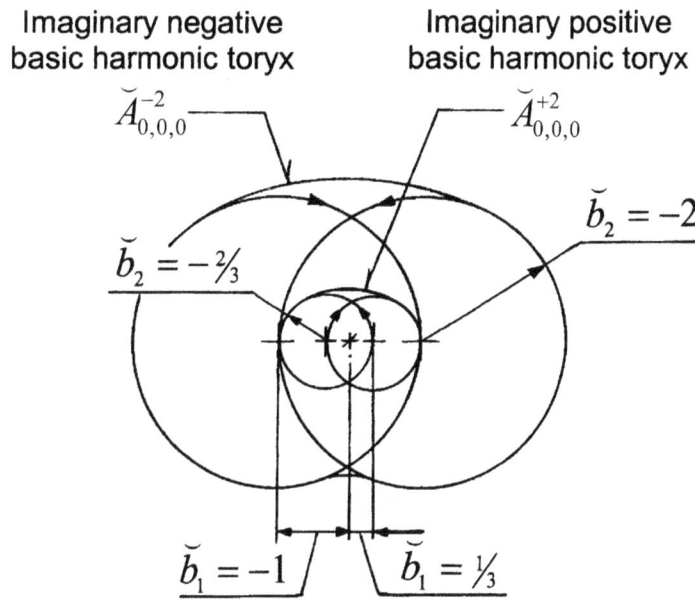

Figure 8.5 Cross-section and dimensions of the vorticity-polarized harmonic lambda singulatron $\breve{a}^0_{0,0,0}$ and its constituent toryces.

Table 8.5. Composition and properties of the vorticity-polarized harmonic lambda singulatron $\breve{a}^0_{0,0,0}$ and its constituent toryces.

Toryx	b_1	$\beta_1 = \beta_{2t}$	w_2	e_t/e	μ_t/μ_N	m_{tg}/m_e	I	N
$\breve{A}^{-2}_{0,0,0}$	-1.0	- 1.732051i	- 0.577350i	-2.0	-23.20784127i	2.000	-6.0	1
$\breve{A}^{+2}_{0,0,0}$	$\frac{1}{3}$	1.732051i	0.577350i	+2.0	+7.73594709i	2.000	$+\frac{2}{3}$	1
Harmonic lambda singulatron $\breve{a}^0_{0,0,0}$				**0.00**	**-7.73594709**	**2.000**	$-\frac{8}{3}$	**1**

8.6 *Self-polarized unexcited harmonic electron* $ae_{H,0,0}^{-1}$ is made up of one self-polarized real negative harmonic toryx $A_{H,0,0}^{-1/2}$ and one self-polarized real negative harmonic toryx $E_{H,0,0}^{-1/2}$ as given by the equation:

$$ae_{H,0,0}^{-1} = A_{H,0,0}^{-1/2} + E_{H,0,0}^{-1/2}$$

Figure 8.6 and Table 8.6 show cross-section, dimensions, compositions and properties of the unexcited harmonic electron $ae_{H,0,0}^{-1}$.

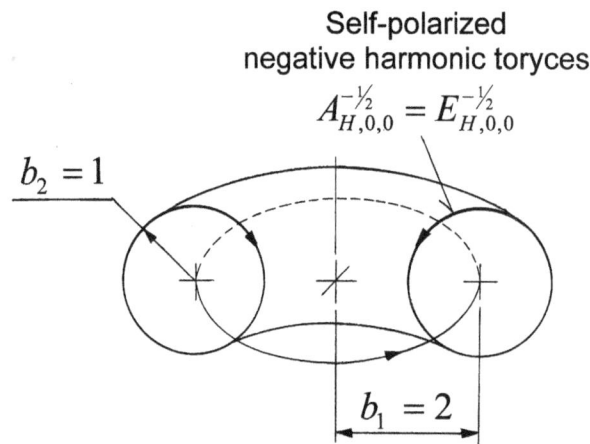

Figure 8.6. Cross-section and dimensions of the self-polarized unexcited harmonic electron $ae_{H,0,0}^{-1}$ and its constituent toryces.

Table 8.6. Composition and properties of the self-polarized unexcited harmonic electron $ae_{H,0,0}^{-1}$ and its constituent toryces.

Toryx	b_1	$\beta_1 = \beta_{2t}$	w_2	e_t/e	μ_t/μ_N	m_{tg}/m_e	N
$A_{H,0,0}^{-1/2}$	2.0	0.866025	1.154701	-0.50	-5.80196032	0.500	1
$E_{H,0,0}^{-1/2}$	2.0	0.866025	1.154701	-0.50	-5.80196032	0.500	1
Self-polarized harmonic electron $ae_{H,0,0}^{-1}$				**-1.00**	**-11.60392064**	**1.000**	**1**

8.7 Self-polarized unexcited harmonic positron $ae_{H,0,0}^{+1}$ is made up of one self-polarized real positive harmonic toryx $A_{H,0,0}^{+1/2}$ and one self-polarized positive harmonic toryx $E_{H,0,0}^{+1/2}$ as the equation:

$$ae_{H,0,0}^{+1} = A_{H,0,0}^{+1/2} + E_{H,0,0}^{+1/2}$$

Figure 8.7 and Table 8.7 show cross-section, dimensions, compositions and properties of the unexcited harmonic positron $ae_{H,0,0}^{+1}$.

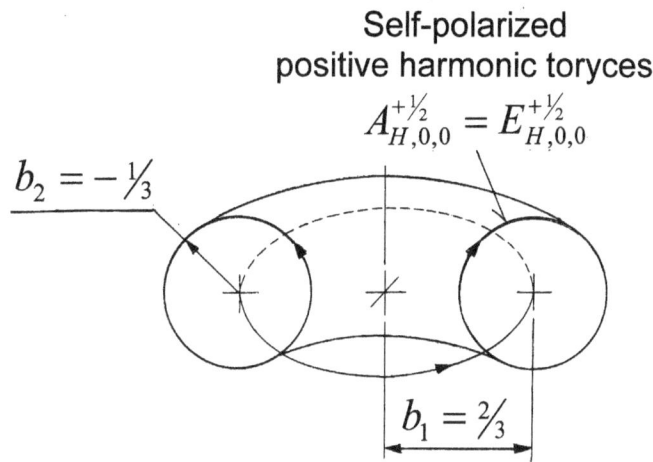

Figure 8.7. Cross-section and dimensions of the self-polarized unexcited harmonic positron $ae_{H,0,0}^{+1}$ and its constituent toryces.

Table 8.7. Composition and properties of the self-polarized unexcited harmonic positron $ae_{H,0,0}^{+1}$ and its constituent toryces.

Toryx	b_1	$\beta_1 = \beta_{2t}$	w_2	e_t/e	μ_t/μ_N	m_{tg}/m_e	N
$A_{H,0,0}^{+1/2}$	$2/3$	0.866025	1.154701	+0.50	+1.93398677	0.500	1
$E_{H,0,0}^{+1/2}$	$2/3$	0.866025	1.154701	+0.50	+1.93398677	0.500	1
Self-polarized harmonic positron $ae_{H,0,0}^{+1}$				+1.00	+ 3.86797355	1.000	1

8.8 Self-polarized excited harmonic electron $ae^{-1}_{H,1,0}$ is made up of one self-polarized real negative excited harmonic toryx $A^{-1/3}_{H,1,0}$ and one self-polarized real negative excited harmonic toryx $E^{-2/3}_{H,1,0}$ as given by the equation:

$$ae^{-1}_{H,1,0} = A^{-1/3}_{H,1,0} + E^{-2/3}_{H,1,0}$$

Figure 8.8 and Table 8.8 show cross-section, dimensions, compositions and properties of the unexcited harmonic electron $ae^{-1}_{H,1,0}$.

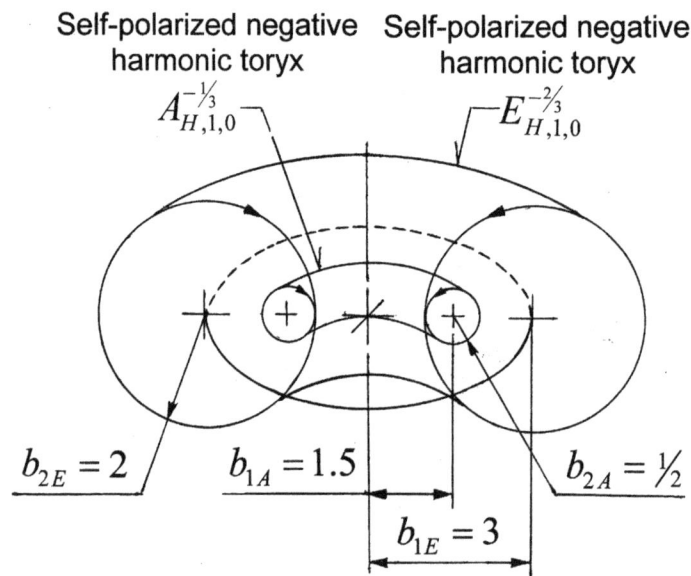

Figure 8.8. Cross-section and dimensions of the self-polarized excited harmonic electron $ae^{-1}_{H,1,0}$ and its constituent toryces.

Table 8.8. Composition and properties of the self-polarized real negative excited harmonic electron $ae^{-1}_{H,1,0}$ and its constituent toryces.

Toryx	b_1	$\beta_1 = \beta_{2t}$	w_2	e_t/e	μ_t/μ_N	m_{tg}/m_e	N
$A^{-1/3}_{H,1,0}$	1.5	0.942809	1.060660i	-1/3	-3.15818717	1/3	1
$E^{-2/3}_{H,1,0}$	3.0	0.745356	1.341641	-2/3	-9.98706475	2/3	1
Self-polarized excited electron $ae^{-1}_{H,1,0}$				**-1.00**	**-13.14525192**	**1.000**	**1**

8.9 Self-polarized excited harmonic positron $ae_{H,1,0}^{+1}$ is made up of one real self-polarized positive excited harmonic toryx $A_{H,1,0}^{+1/3}$ and one self-polarized real positive excited harmonic toryx $E_{H,1,0}^{+2/3}$ as given by the equation:

$$ae_{H,1,0}^{+1} = A_{H,1,0}^{+1/3} + E_{H,1,0}^{+2/3}$$

Figure 8.9 and Table 8.9 show cross-section, dimensions, compositions and properties of the unexcited harmonic positron $ae_{H,1,0}^{+1}$.

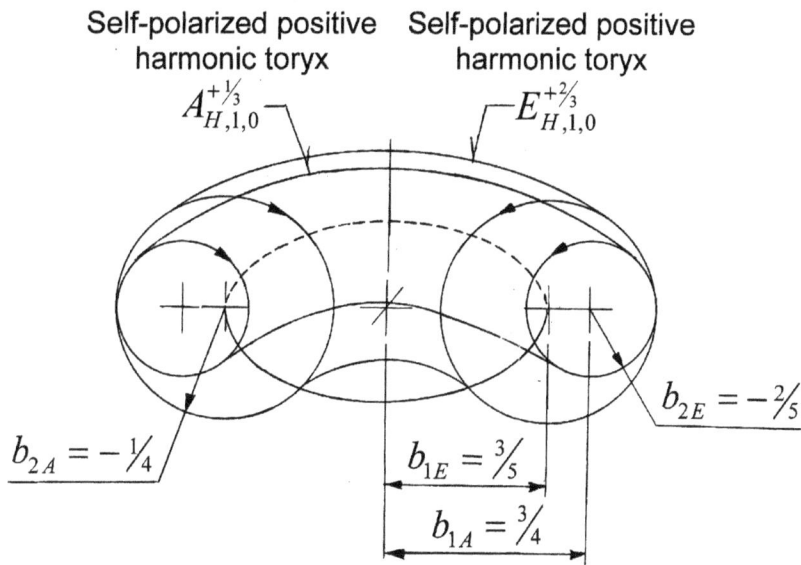

Figure 8.9. Cross-section and dimensions of the self-polarized excited harmonic positron $ae_{H,1,0}^{+1}$ and its constituent toryces.

Table 8.9. Composition and properties of the self-polarized excited harmonic positron $ae_{H,1,0}^{+1}$ and its constituent toryces.

Toryx	b_1	$\beta_1 = \beta_{2t}$	w_2	e_t/e	μ_t/μ_N	m_{tg}/m_e	N
$A_{H,1,0}^{+1/3}$	$3/4$	0.942809	1.060660	$+1/3$	+1.57909359	$1/3$	1
$E_{H,1,0}^{+2/3}$	$3/5$	0.745356	1.341641	$+2/3$	+1.99741295	$2/3$	1
Self-polarized excited positron $ae_{H,1,0}^{+1}$				**+1.00**	**+3.57650654**	**1.000**	**1**

8.10 *Self-polarized unexcited golden electron* $ae_{G,0,0}^{-1}$ is made up of one self-polarized real negative golden toryx $A_{G,0,0}^{-1/2}$ and one self-polarized real negative golden toryx $E_{G,0,0}^{-1/2}$ as given by the equation:

$$ae_{G,0,0}^{-1} = A_{G,0,0}^{-1/2} + E_{G,0,0}^{-1/2}$$

Figure 8.10 and Table 8.10 show cross-section, dimensions, compositions and properties of the self-polarized unexcited golden electron $ae_{G,0,0}^{-1}$.

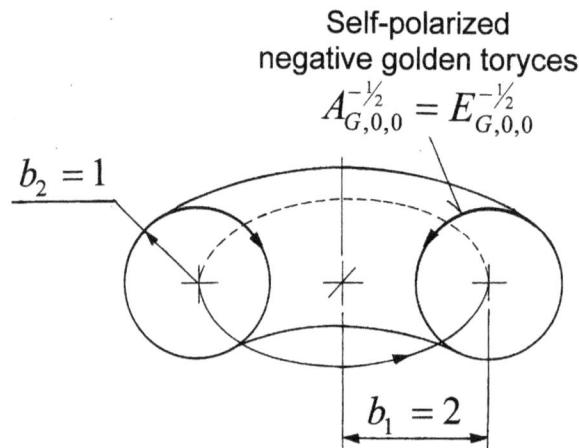

Figure 8.10. Cross-section and dimensions of the self-polarized unexcited golden electron $ae_{G,0,0}^{-1}$ and its constituent toryces.

Table 8.10. Composition and properties of the self-polarized unexcited golden electron $ae_{G,0,0}^{-1}$ and its constituent toryces.

Toryx	b_1	$\beta_1 = \beta_{2t}$	w_2	e_t/e	μ_t/μ_N	m_{tg}/m_e	N
$A_{G,0,0}^{-1/2}$	2.0	0.866025	1.154701	-0.50	-5.80196032	0.500	1
$E_{G,0,0}^{-1/2}$	2.0	0.866025	1.154701	-0.50	-5.80196032	0.500	1
Self-polarized golden electron $ae_{G,0,0}^{-1}$				**-1.00**	**-11.60392064**	**1.000**	**1**

8.11 *Self-polarized unexcited golden positron* $ae_{G,0,0}^{+1}$ is made up of one self-polarized real positive harmonic toryx $A_{G,0,0}^{+\frac{1}{2}}$ and one self-polarized real positive harmonic toryx $E_{G,0,0}^{+\frac{1}{2}}$ as given by the equation:

$$ae_{G,0,0}^{+1} = A_{G,0,0}^{+\frac{1}{2}} + E_{G,0,0}^{+\frac{1}{2}}$$

Figure 8.11 and Table 8.11 show cross-section, dimensions, compositions and properties of the self-polarized unexcited golden positron $ae_{G,0,0}^{+1}$.

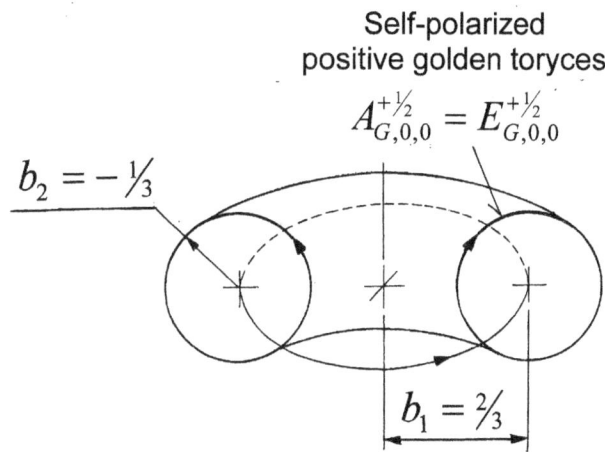

Figure 8.11. Cross-section and dimensions of the self-polarized unexcited golden positron $ae_{G,0,0}^{+1}$ and its constituent toryces.

Table 8.11. Composition and properties of the of the self-polarized unexcited golden positron $ae_{G,0,0}^{+1}$ and its constituent toryces.

Toryx	b_1	$\beta_1 = \beta_{2t}$	w_2	e_t/e	μ_t/μ_N	m_{tg}/m_e	N
$A_{G,0,0}^{+\frac{1}{2}}$	$\frac{2}{3}$	0.866025	1.154701	+0.50	+1.93398677	0.500	1
$E_{G,0,0}^{+\frac{1}{2}}$	$\frac{2}{3}$	0.866025	1.154701	+0.50	+1.93398677	0.500	1
Self-polarized golden positron $ae_{G,0,0}^{+1}$				+1.00	+ 3.86797355	1.000	1

8.12 *Self-polarized excited golden electron* $ae_{G,1,0}^{-1}$ is made up of one self-polarized real negative excited golden toryx $A_{G,1,0}^{-\phi^{-2}}$ and one self-polarized real negative excited golden toryx $E_{G,1,0}^{-\phi^{-1}}$ as by the equation:

$$ae_{G,1,0}^{-1} = A_{G,1,0}^{-\phi^{-2}} + E_{G,1,0}^{-\phi^{-1}}$$

Figure 8.12 and Table 8.12 show cross-section, dimensions, compositions and properties of the self-polarized excited golden electron $ae_{G,1,0}^{-1}$.

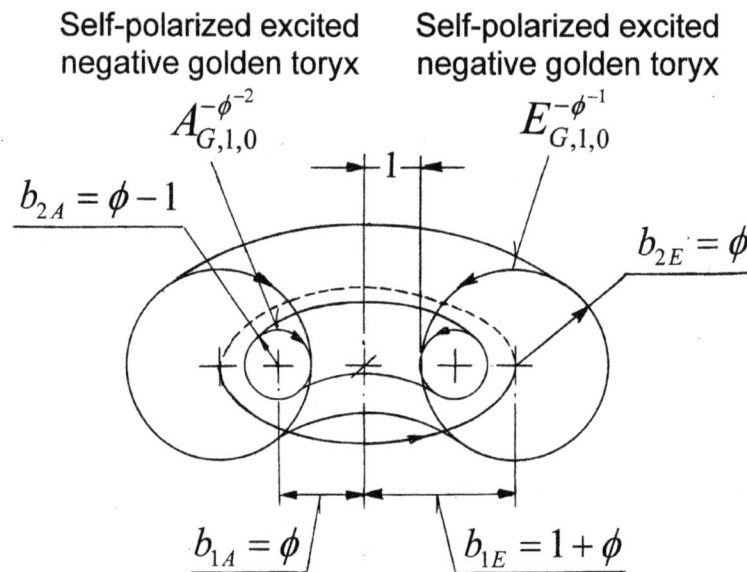

Figure 8.12. Cross-section and dimensions of the self-polarized excited golden electron $ae_{G,1,0}^{-1}$ and its constituent toryces.

Table 8.12. Composition and properties of the self-polarized excited golden electron $ae_{G,1,0}^{-1}$ and its constituent toryces.

Toryx	b_1	$\beta_1 = \beta_{2t}$	w_2	e_t/e	μ_t/μ_N	m_{tg}/m_e	N
$A_{G,1,0}^{-\phi^{-2}}$	ϕ	0.924176	1.082045	$-\phi^{-2}$	- 3.8265848	ϕ^{-2}	1
$E_{G,1,0}^{-\phi^{-1}}$	$1+\phi$	0.786151	1.272020	$-\phi^{-1}$	- 8.5219296	ϕ^{-1}	1
Self-polarized excited golden electron $ae_{G,1,0}^{-1}$				**-1.00**	**-12.3485144**	**1.000**	**1**

8.13 *Self-polarized excited golden positron* $ae_{G,1,0}^{+1}$ is made up of one self-polarized real positive excited golden toryx $A_{G,1,0}^{+\phi^{-2}}$ and one self-polarized real positive excited golden toryx $E_{G,1,0}^{+\phi^{-1}}$ as given by the equation:

$$ae_{G,1,0}^{+1} = A_{G,1,0}^{+\phi^{-2}} + E_{G,1,0}^{+\phi^{-1}}$$

Figure 8.13 and Table 8.13 show cross-section, dimensions, compositions and properties of the self-polarized excited golden electron $ae_{G,1,0}^{-1}$.

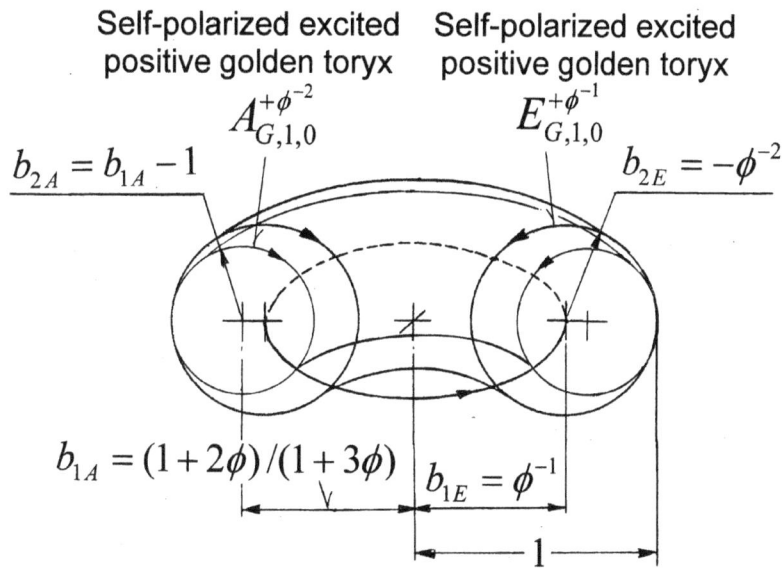

Figure 8.13. Cross-section and dimensions of the self-polarized excited golden positron $ae_{G,1,0}^{+1}$ and its constituent toryces.

Table 8.13. Composition and properties of the self-polarized excited golden positron $ae_{G,1,0}^{+1}$ and its constituent toryces.

Toryx	b_1	$\beta_1 = \beta_{2t}$	w_2	e_t/e	μ_t/μ_N	m_{tg}/m_e	N
$A_{G,1,0}^{+\phi^{-2}}$	$\dfrac{1+2\phi}{1+3\phi}$	0.924176	1.082045	$+\phi^{-2}$	+1.7113008	ϕ^{-2}	1
$E_{G,1,0}^{+\phi^{-1}}$	ϕ^{-1}	0.786151	1.272020	$+\phi^{-1}$	+2.0117547	ϕ^{-1}	1
Self-polarized excited golden positron $ae_{G,1,0}^{+1}$				+1.00	+3.7230554	1.000	1

8.14 *Excited lambda electron* $e_{m,n,q}^{-1}$ is made up of one excited real negative lambda toryx $E_{m,n,q}^{-}$ and one excited imaginary negative lambda toryx $\breve{E}_{m,n,q}^{-}$ as shown by the equation:

$$e_{m,n,q}^{-1} = E_{m,n,q}^{-} + \breve{E}_{m,n,q}^{-}$$

Figure 8.14 and Table 8.14 show cross-section, dimensions, compositions and properties of the excited lambda electron $e_{2,1,0}^{-1}$ corresponding to the spacetime level of atoms *L2* (ordinary matter) see Table 8.2.2.

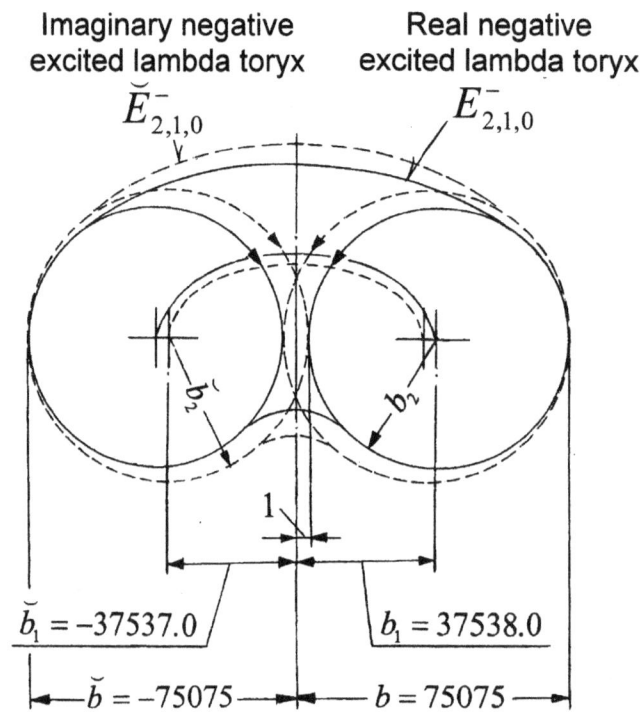

Figure 8.14. Cross-section and dimensions of the excited lambda electron $e_{2,1,0}^{-1}$ of ordinary matter.

Table 8.14. Composition and properties of the excited lambda electron $e_{2,1,0}^{-1}$ of ordinary matter.

Toryx	b_1	e_t/e	μ_t/μ_N	m_{tg}/m_e	I	N
$E_{2,1,0}^{-}$	37538.0	-0.99997336	-1835.609190	0.99997336	-75075.0	1.00002664
$\breve{E}_{2,1,0}^{-}$	-37538.0	-1.00002664	-1835.706994i	1.00002664	+75075.0	0.99997336
Electron $e_{2,1,0}^{-1}$	- 1.0000000	-1835.658091	1.00000000	0.000	1

8.15 Excited lambda positron $e_{m,n,q}^{+1}$ is made up of one excited real positive lambda toryx $E_{m,n,q}^{+}$ and one excited imaginary positive lambda toryx $\breve{E}_{m,n,q}^{+}$ as shown by the equation:

$$e_{m,n,q}^{+1} = E_{m,n,q}^{+} + \breve{E}_{m,n,q}^{+}$$

Figure 8.15 and Table 8.15 show cross-section, dimensions, compositions and properties of the excited lambda positron $e_{2,1,0}^{+1}$ of ordinary matter.

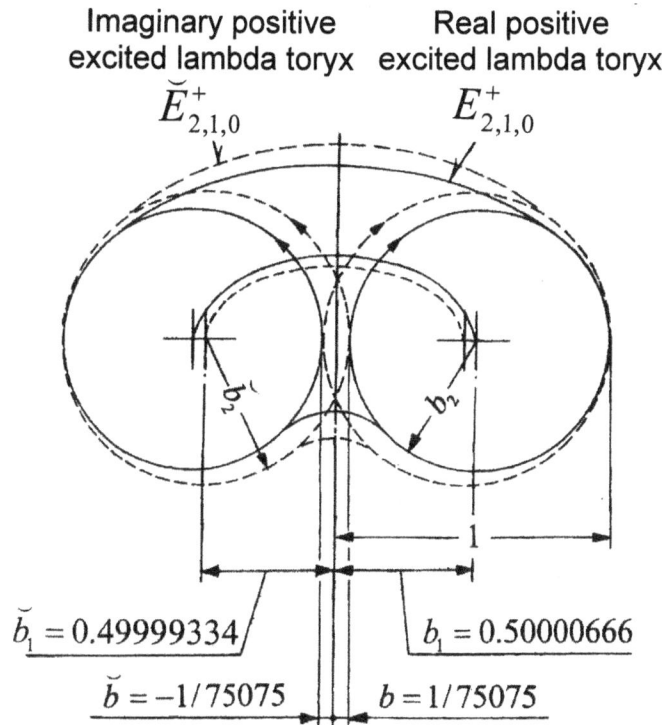

Figure 8.15. Cross-section and dimensions of the excited lambda positron $e_{2,1,0}^{+1}$ of ordinary matter.

Table 8.15. Composition and properties of the excited lambda positron $e_{2,1,0}^{+1}$ of ordinary matter.

Toryx	b_1	e_t/e	μ_t/μ_N	m_{tg}/m_e	I	N
$E_{2,1,0}^{+}$	0.50000666	+0.99997336	+0.024450	0.99997336	$+1.3320\times10^{-5}$	1.00002664
$\breve{E}_{2,1,0}^{+}$	0.49999334	+1.00002664	+0.024452i	1.00002664	-1.3320×10^{-5}	0.99997336
Positron $e_{2,1,0}^{+1}$	+1.0000000	+0.024451	1.00000000	0.000	1	

8.16 *Excited lambda ethertron* $a^0_{m,n,q}$ is made up of one excited real negative lambda toryx $A^-_{m,n,q}$ and one excited real positive lambda toryx $A^+_{m,n,q}$ as shown by the equation:

$$a^0_{m,n,q} = A^-_{m,n,q} + A^+_{m,n,q}$$

Figure 8.16 and Table 8.16 show cross-section, dimensions, compositions and properties of the excited lambda ethertron $a^0_{1,1,0}$ of ordinary matter.

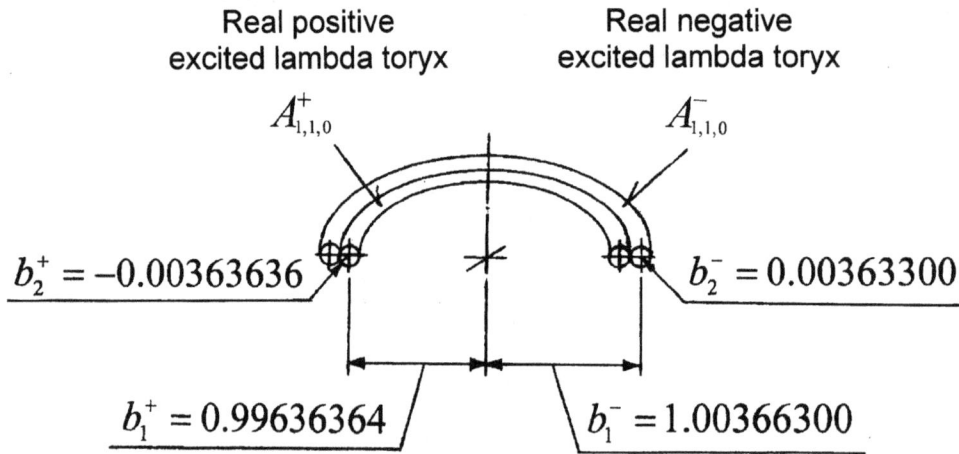

Figure 8.16. Cross-section and dimensions of the excited lambda ethertron $a^0_{1,1,0}$ of ordinary matter.

Table 8.16. Composition and properties of the excited lambda ethertron $a^0_{1,1,0}$ of ordinary matter.

Toryx	b_1	e_t/e	μ_t/μ_N	m_{tg}/m_e	I	N
$A^-_{1,1,0}$	1.0036630	- 0.003650	-0.02454023	0.00364964	+0.003676	1
$A^+_{1,1,0}$	0.9963636	+ 0.003650	+0.02436175	0.00364964	-0.003623	1
Ethertron $a^0_{1,1,0}$	**0.00**	**-0.00008924**	**0.00364964**	**+0.000027**	**1**

8.17 Excited lambda singulatron $\breve{a}^0_{m,n,q}$ is made up of one excited imaginary negative lambda toryx $\breve{A}^-_{m,n,q}$ and one excited imaginary positive lambda toryx $\breve{A}^+_{m,n,q}$ as shown by the equation:

$$\breve{a}^0_{m,n,q} = \breve{A}^-_{m,n,q} + \breve{A}^+_{m,n,q}$$

Figure 8.17 and Table 8.17 show cross-section, dimensions, compositions and properties of the excited lambda singulatron $\breve{a}^0_{1,1,0}$ of ordinary matter.

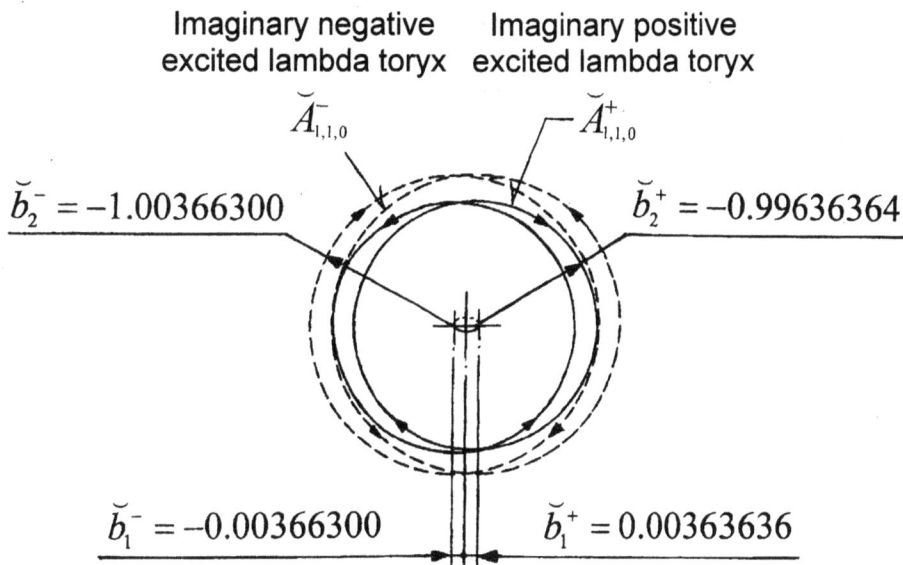

Imaginary negative
excited lambda toryx
$\breve{A}^-_{1,1,0}$

Imaginary positive
excited lambda toryx
$\breve{A}^+_{1,1,0}$

$\breve{b}^-_2 = -1.00366300$

$\breve{b}^+_2 = -0.99636364$

$\breve{b}^-_1 = -0.00366300$

$\breve{b}^+_1 = 0.00363636$

Figure 8.17. Cross-section and dimensions of the excited lambda singulatron $\breve{a}^0_{1,1,0}$ of ordinary matter.

Table 8.17. Properties of the excited lambda singulatron $\breve{a}^0_{1,1,0}$ of ordinary matter.

Toryx	b_1	e_t/e	μ_t/μ_N	m_{tg}/m_e	I	N
$\breve{A}^-_{1,1,0}$	0.00366300	- 273.998175	-1842.382113i	274.00	-276.007326	1
$\breve{A}^+_{1,1,0}$	0.00363636	+ 273.998175	+1828.982971i	274.00	+272.007273	1
Singulatron $\breve{a}^0_{1,1,0}$	**0.00**	- 6.69957132		**274.00**	-2.000027	**1**

8.18 *Combined harmonic lambda ce-tron* $ce_{0,0,0}^{0}$ is made up of one harmonic lambda electron $e_{0,0,0}^{-1}$ and one harmonic lambda positron $e_{0,0,0}^{+1}$ as shown by the equation:

$$ce_{0,0,0}^{0} = e_{0,0,0}^{-1} + e_{0,0,0}^{+1}$$

Figure 8.18 and Table 8.18 show cross-section, dimensions, compositions and properties of the combined harmonic lambda ce-tron $ce_{0,0,0}^{0}$.

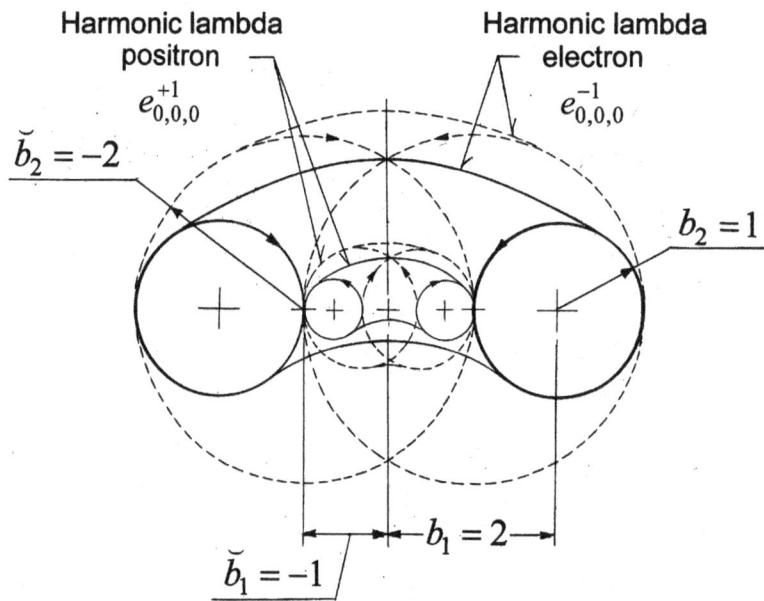

Figure 8.18. Cross-section and dimensions of properties of the combined harmonic lambda ce-tron $ce_{0,0,0}^{0}$ and its constituent toryces.

Table 8.18. Composition and properties of properties of the combined harmonic lambda ce-tron $ce_{0,0,0}^{0}$ and its constituent toryces.

Harmonic lambda trons		e_t/e	μ_t/μ_N	m_{tg}/m_e	I	N
Electron $e_{0,0,0}^{-1}$	Table 8.1	-1.00	-11.60392064	1.000	0.00	1
Positron $e_{0,0,0}^{+1}$	Table 8.3	+1.00	+3.86797355	1.000	0.00	1
Combined harmonic ce-tron $ce_{0,0,0}^{0}$		**0.00**	**-7.73594709**	**2.000**	**0.00**	**1**

8.19 *Combined harmonic lambda ca-tron* $ca^0_{0,0,0}$ is made up of one harmonic lambda singulatron $\breve{a}^0_{0,0,0}$ and four harmonic lambda ethertrons $a^0_{0,0,0}$ as shown by the equation:

$$ca^0_{0,0,0} = \breve{a}^0_{0,0,0} + 4a^0_{0,0,0}$$

Figure 8.19 and Table 8.19 show cross-section, dimensions, compositions and properties of the combined harmonic lambda ca-tron $ca^0_{0,0,0}$.

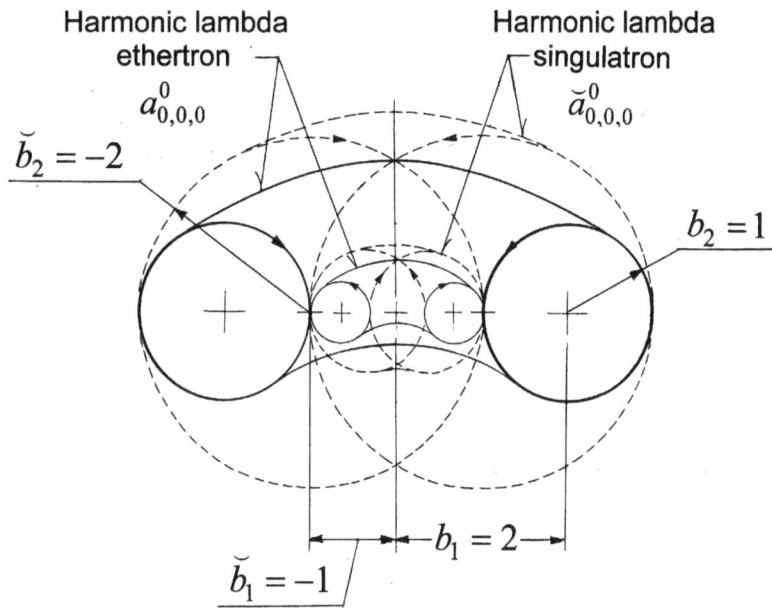

Figure 8.19. Cross-section and dimensions of properties of the combined harmonic lambda ca-tron $ca^0_{0,0,0}$ and its constituent toryces.

Table 8.19. Composition and properties of properties of the combined harmonic lambda ca-tron $ca^0_{0,0,0}$ and its constituent toryces.

Harmonic lambda trons		e_t/e	μ_t/μ_N	m_{tg}/m_e	I	N
Singulatron $\breve{a}^0_{0,0,0}$	Table 8.5	0.00	-7.73594709	2.000	$-8/3$	1
Ethertron $a^0_{0,0,0}$	Table 8.4	0.00	-1.93398677	0.500	$+2/3$	4
Combined harmonic ca-tron $ca^0_{0,0,0}$		**0.00**	**-7.73594709**	**2.000**	**0.00**	**1**

8.20 Combined excited lambda ce-tron $ce^0_{m,n,q}$ is made up one electron $e^{-1}_{m,n,q}$ and one positron $e^{+1}_{m,n,q}$ as shown by the equation and in Table 8.20:

$$ce^0_{m,n,q} = e^{-1}_{m,n,q} + e^{+1}_{m,n,q}$$

Table 8.20. Composition and properties of the combined excited lambda ce-tron $e^{-1}_{2,1,0}$ of ordinary matter.

Excited lambda trons		e_t/e	μ_t/μ_N	m_{tg}/m_e	I	N
Electron $e^{-1}_{2,1,0}$	Table 8.14	- 1.000000	-1835.658091	1.000000	0.0	1
Positron $e^{+1}_{2,1,0}$	Table 8.15	+1.000000	+0.024451	1.000000	0.0	1
Combined ce-tron $ce^0_{2,1,0}$		**0.000000**	**-1835.63364**	**2.000000**	**0.0**	**1**

8.21 Combined excited lambda ca-tron $ca^0_{m,n,q}$ is made up one singulatron $\breve{a}^0_{m,n,q}$ and N ether-trons $a^0_{m,n,q}$ as shown by the equation and in Table 8.21:

$$ca^0_{m,n,q} = \breve{a}^0_{m,n,q} + N a^0_{m,n,q}$$

Table 8.21. Composition and properties of the combined excited lambda ca-tron $ca^0_{m,n,q}$ of ordinary matter.

Excited lambda trons		e_t/e	μ_t/μ_N	m_{tg}/m_e	I	N
Singulatron $\breve{a}^0_{1,1,0}$	Table 8.17	0.00	- 6.69957132	274.000000	-2.000027	1
Ethertron $a^0_{1,1,0}$	Table 8.16	0.00	-0.00008924	0.00364964	+0.000027	75076
Combined excited ca-tron $ca^0_{1,1,0}$	**0.000000**	**0.00000000**	**274.000000**	**0.000000**	**1**

For a case in which the excitation quantum states n of the toryces forming a singulatron and N ethertrons correspond to the spacetime limits, the combined excited ca-tron $ca^0_{1,1,0}$ becomes the extreme form of **ether** (Fig. 8.21) in which one singulatron is surrounded by N ethertrons with N defined by the equation:

$$N = \left(\frac{1 \pm b_s}{b_s}\right)^2 \approx 1.511267 \times 10^{51}$$

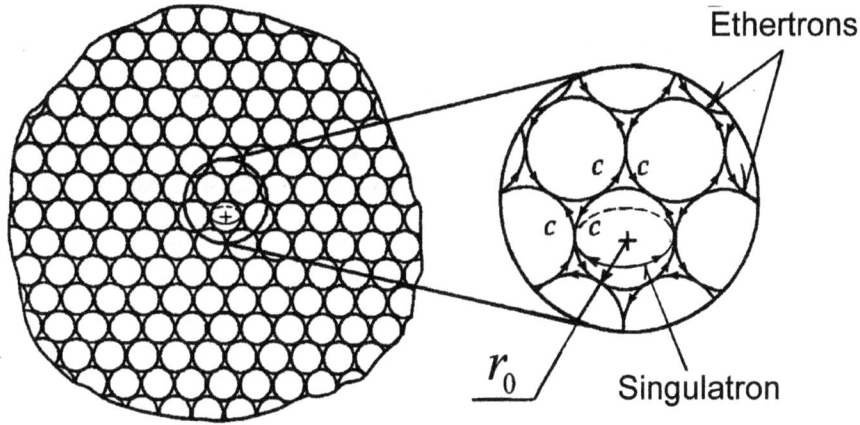

Figure 8.21. Extreme case of ether.

8.22 Oscillated excited lambda electrons – Figure 8.22 shows a plot of a natural logarithm of the toryx oscillation factor $Q_q = m_{tg} / m_e$ as a function of the toryx oscillation quantum states q calculated from Eq. (6.4-1).

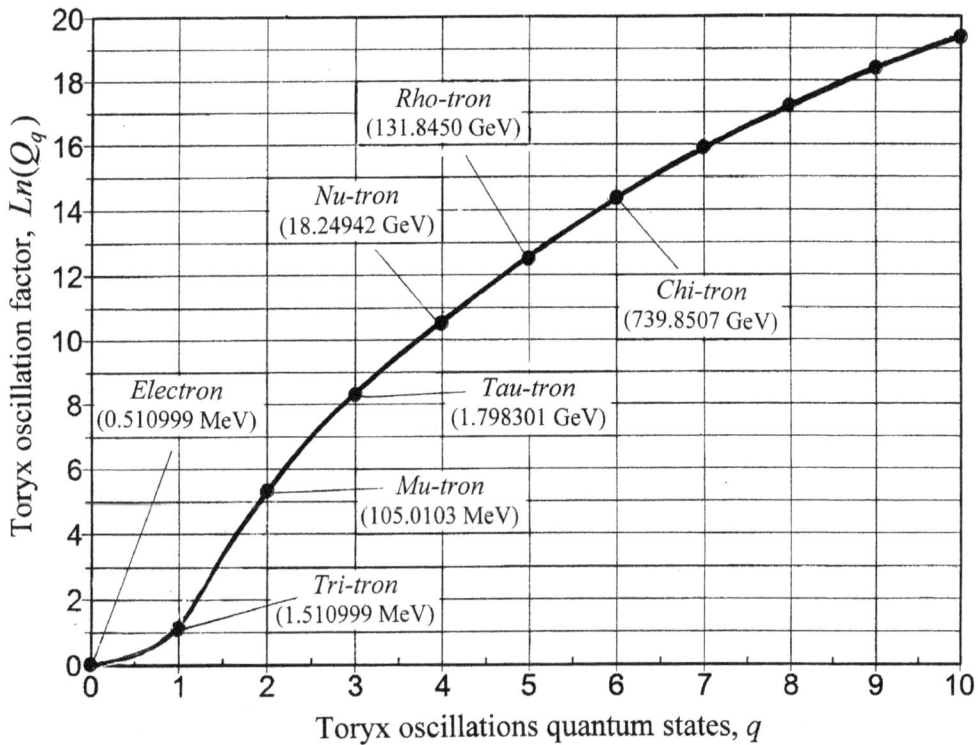

Figure 8.22. Toryx oscillation factor Q_q as a function of the toryx oscillation quantum states q.

Table 8.22 compares calculated physical properties of the oscillated excited lambda electrons with measured physical properties of leptons. The relationship between the toryx oscillation factor Q_q and the toryx oscillation quantum states q shown in Table 8.22 are defined by Eq. (6.4-1).

Table 8.22. Calculated & measured properties of oscillated excited lambda electrons.

Oscillated excited electrons		Physical properties		
Name	Symbol	μ_t / μ_B	$m_{tg} / m_e = Q_q$	GeV/c^2
Electron	$e_{2,1,0}^{-1}$	**-0.99973052**	**1.000000**	**0.00051099893**
Measured values:		-1.00115965	1.000000	0.00051099893
Calc./measured ratio:		0.9986	1.0000	1.00
$3e-tron$	$e_{2,1,1}^{-1}$	**-0.33324351**	**3.000000**	**0.00151099893**
Measured values:		?	?	0.00151099893
Calc./measured ratio:		-0.004841970	?	1.00
$\mu-tron$	$e_{2,1,2}^{-1}$	**-4.864869 × 10^{-3}**	**205.500000**	**0.1050102797**
Measured values		-4.841970 × 10^{-3}	206.768284	0.1056583668
Calc./measured ratio		1.0047	0.9939	0.9939
$\tau-tron$	$e_{2,1,3}^{-1}$	**-2.840799 × 10^{-4}**	**3519.18750**	**1.798301041**
Measured values		?	3478.18283	1.776990000
Calc./measured ratio		?	1.0120	1.0120
$\nu-tron$	$e_{2,1,4}^{-1}$	**-2.799328 × 10^{-5}**	**35713.23612**	**18.2494254**
Measured values		?	?	?
Calc./measured ratio		?	?	?
$\rho-tron$	$e_{2,1,5}^{-1}$	**-3.874712 × 10^{-6}**	**258014.1805**	**131.8449697**
Measured values		?	244795.0427	125.0900000
Calc./measured ratio		?	1.0540	1.0540
$\chi-tron$	$e_{2,1,6}^{-1}$	**-6.904923 × 10^{-7}**	**1447851.734**	**739.8506841**
Measured values		?	1468.713450	750 ?
Calc./measured ratio		?	0.98646758	0.98646758

It is follows from Table 8.22 and Figure 8.22:

- Calculated masses of **electron, mu-tron,** and **tau-tron** are very close to their respective measured values.
- Calculated mass of **rho-tron** is very close to the mass of Higgs boson of about 125 GeV measured at CERN in 2017.

8.23 *Reality-polarized resonant oscillated harmonic lambda electron* $er_{0,0,1}^{-1}$ is made up of one real negative oscillated harmonic toryx $E_{0,0,1}^{-1/2}$ and one imaginary negative resonant oscillated harmonic toryx $\breve{E}r_{0,0,1}^{-3/2}$ as shown by the equation:

$$er_{0,0,1}^{-1} = E_{0,0,1}^{-1/2} + \breve{E}r_{0,0,1}^{-3/2}$$

Figure 8.23 and Table 8.23 show cross-section, dimensions, compositions and properties of the reality-polarized oscillated resonant harmonic lambda electron $er_{0,0,1}^{-1}$.

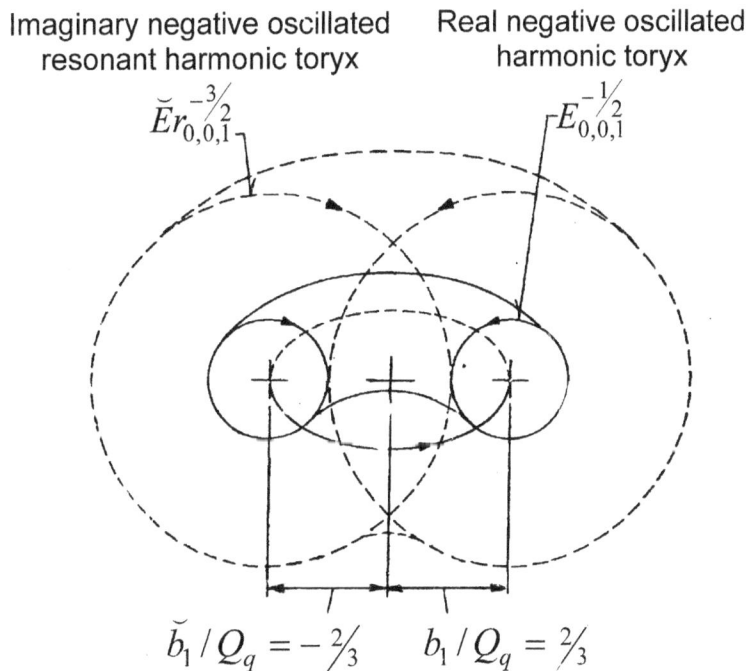

Figure 8.23. Cross-section and dimensions of the reality-polarized resonant oscillated harmonic lambda electron $er_{0,0,1}^{-1}$ and its constituent toryces.

Table 8.23. Composition and properties of the reality-polarized resonant oscillated harmonic lambda electron $er_{0,0,1}^{-1}$ and its constituent toryces.

$er_{0,0,1}^{-1}$ and its constituent toryces.

Trons	Tables	b_1	$\beta_1 = \beta_{2t}$	w_2	e_t/e	μ_t/μ_N	m_{tg}/m_e	I	N
$E_{0,0,1}^{-1/2}$	8.2	2.0	0.866025	1.154701	-0.50	- 1.93398677	1.500	+4.5	1
$\breve{E}r_{0,0,1}^{-3/2}$	8.2	-2.0	-1.118034i	- 0.894427	-1.50	- 7.49029856	4.500	-22.5	1
Resonant oscillated harmonic electron $er_{0,0,1}^{-1}$					**-1.00**	**-4.71214267**	**3.000**	**-9.00**	**1**

9. NUCLEONS, LIGHT ATOMS & ISOTOPES

Nucleons are made up of three main parts: *nucleon crystal*, *nucleon core* and *nucleon leptons*, all of them contained inside a spherical membrane.

9.1 Nucleon Crystal Structure

In a nucleon, the degrees of freedom of its constituent elementary trons are reduced because they are located inside the spherical spaces (Fig. 9.1.1) located around the center 1 and the vertices 2 through 9 of a nucleon bi-pyramid hexagonal crystal structure called the *nucleon crystal*.

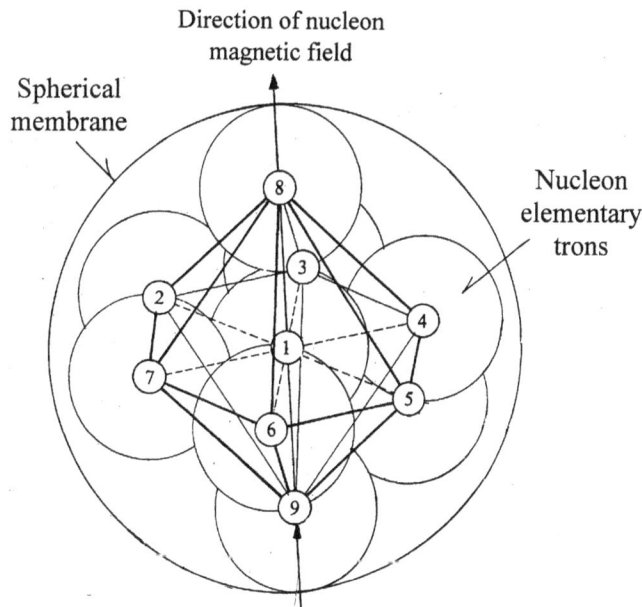

Figure 9.1.1. Isometric view of a nucleon crystal.

As shown in Figs. 9.1.2 and 9.1.3, there are two parts in the nucleon crystal, outer and inner. The inner part is three times smaller than the outer part and it is located inside the outer part. Each part has the center 1 and eight vertices 2 through 9. The vertices of the outer part (Fig. 9.1.3) are formed at the intersections of leading string strings a, b, c and d of real negative harmonic toryces, while the vertices of the inner part are formed at the intersections of leading string strings a, b, c and d of real positive harmonic toryces. The leading strings a, b, c and d of both real negative and positive harmonic toryces are located in the planes A, B, C and D with the centers of leading strings located at the crystal center 1.

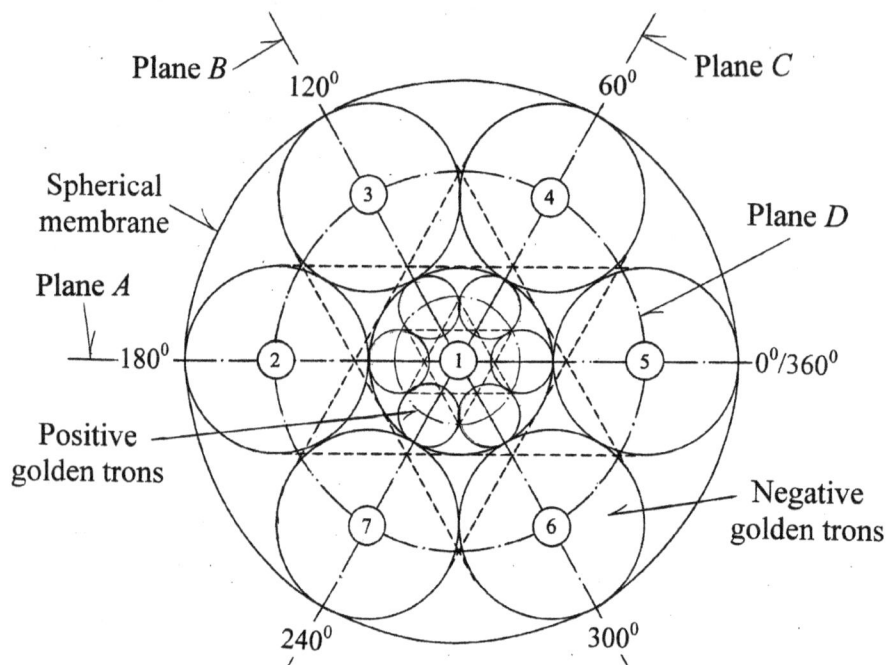

Figure 9.1.2. Cross-section of outer and inner parts of a nucleon crystal.

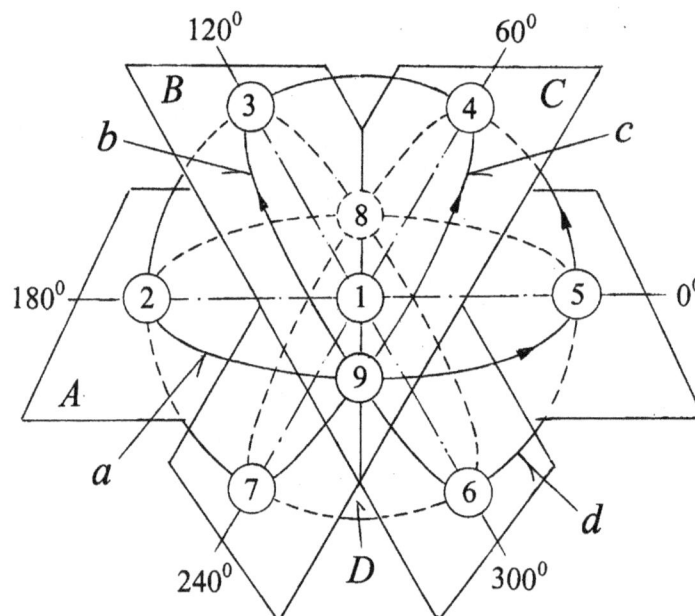

Figure 9.1.3. Isometric view of either outer or inner part of a nucleon crystal.

Table 9.1 shows compositions and properties of the nucleon crystal $\downarrow ncr^0_{0,0,0}$ and its constituent trons that are the same for any matter level L shown in Table 7.4.

The directions of the tron magnetic fields are perpendicular to the related planes A, B, C and D in which they reside. A contribution by each tron to the total magnetic moment of the nucleon crystal is proportional to the sine of the angle γ between the plane in which the tron resides and the nucleon magnetic direction shown in Figs. 9.1.2 and 9.1.3. The singulatron $\downarrow \breve{a}e^0_{0,0,0}$ of the nucleon crystal $\downarrow ncr^0_{0,0,0}$ coexist interchangeably with the ethertrons $\downarrow ae^0_{0,0,0}$ and $\uparrow ae^0_{0,0,0}$. To obey the tron polarization law, each singulatron must be accompanied by four ethertrons as described by Eq. (7.4-3).

Table 9.1.1. Compositions and properties of the unexcited nucleon crystal $\downarrow ncr^0_{0,0,0}$

Trons	Planes Fig. 9.1.3	Angle γ	μ/μ_N	m_g/m_e	I_t	N
Singulatron $\downarrow \breve{a}e^0_{0,0,0}$	A	90^0	-7.73594709	2.00	$+\frac{8}{3}$	1
Ethertron $\downarrow ae^0_{0,0,0}$	A	90^0	-1.93398677	0.50	$-\frac{2}{3}$	1
Ethertron $\uparrow ae^0_{0,0,0}$.	B	30^0	+0.96699339	0.50	$-\frac{2}{3}$	1
Ethertron $\downarrow ae^0_{0,0,0}$	C	30^0	+0.96699339	0.50	$-\frac{2}{3}$	1
Ethertron $ae^0_{0,0,0}$	D	0^0	0.00	0.50	$-\frac{2}{3}$	1
Nucleon crystal $\downarrow ncr^0_{0,0,0}$			**-3.86797355**	**2.00**	**0.0**	**1**

9.2 Nucleon Core

Nucleon core, or shortly the **nucore**, of any spacetime level L is made up of the singulatrons $\breve{a}^0_{m,n,q}$ and the ethertrons $a^0_{m,n,q}$ that reside inside spheres (see Figs. 9.1.1 – 9.1.3) surrounding the center 1 and the vertices 2 through 7 of the nucleon crystal $\downarrow ncr^0_{m,n,q}$. The vertices 8 and 9 of the nucleon crystal remain vacant to provide binding between adjacent nucleons. The nucore contains a bulk of the nucleon mass, and it requires seven singulatrons to model masses of known nucleons. To obey the tron polarization law, each singulatron must be accompanied by T ethertrons as described by Eq. (7.4-3). Consequently, in the nucore $nco^0_{1,1,0}$ of the spacetime level $L2$ (ordinary matter) the number of ethertrons $T = 75076$ and its composition is described by the equation:

$$\uparrow nco^0_{1,1,0} = 3(\downarrow \breve{a}^0_{1,1,0} + T \downarrow a^0_{1,1,0}) + 4(\uparrow \breve{a}^0_{1,1,0} + T \uparrow a^0_{1,1,0}) \qquad (9.2-1)$$

Table 9.2 shows components and properties of the nucore $nco^0_{1,1,0}$ of the ordinary matter $L2$.

Table 9.2. Components and properties of the nucore $nco^0_{1,1,0}$ of the ordinary matter *L2*.

Excited trons	Locations Fig. 9.1.3	Tables	μ/μ_N	m_g/m_e	I_t	N
Singulatron $\downarrow \breve{a}^0_{1,1,0}$	1, 2, 5	8.17	-6.69957132	274.00	+2.0	3
Singulatron $\uparrow \breve{a}^0_{1,1,0}$	3, 4, 6, 7	8.17	+6.69957132	274.00	+2.0	4
Ethertron $\downarrow a^0_{1,1,0}$	1, 2, 5	8.16	-0.00008924	0.00364964	-2.66×10^{-5}	$3T$
Ethertron $\uparrow a^0_{1,1,0}$	3, 4, 6, 7	8.16	+0.00008924	0.00364964	-2.66×10^{-5}	$4T$
Nucore $nco^0_{1,1,0}$			**+6.69957132**	**1918.0000**	**0.0**	**1**

9.3 Nucleons

Described below are the compositions and properties of nucleons with the same nucleon structure shown in Fig. 9.3. In this figure, small circles indicate locations occupied by the nucleon constituent trons, while empty circles indicate locations occupied by constituent trons of adjacent nucleons in complex nuclei.

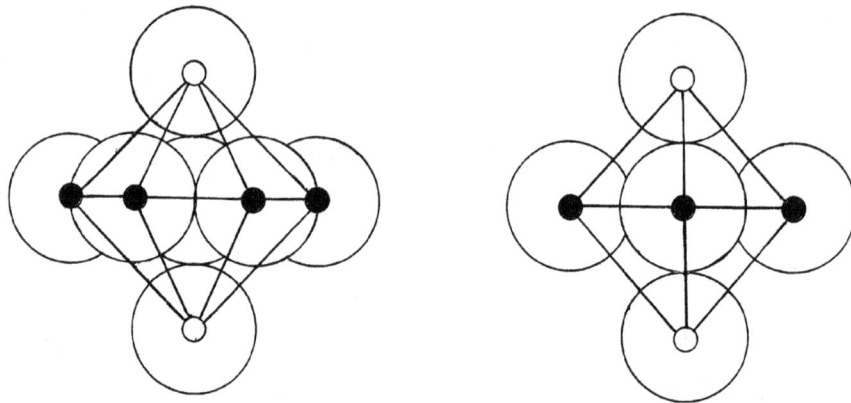

Figure 9.3. Front view (left) and side view (right) of a nucleon crystal structure.

Proton $\uparrow p^{+1}_{L2}$ of the ordinary matter *L2* is made up of one nucleon crystal $\downarrow ncr^0_{0,0,0}$, one nucore $\uparrow nco^0_{1,1,0}$ and one excited lambda positron $\uparrow e^{+1}_{2,1,0}$ as shown as shown in Table 9.3.1 and by the equation:

$$\uparrow p^{+1}_{L2} = \downarrow ncr^0_{0,0,0} + \uparrow nco^0_{1,1,0} + \uparrow e^{+1}_{2,1,0} \tag{9.3-1}$$

Table 9.3.1. Components and properties of the proton $\uparrow p_{L2}^{+1}$ of the ordinary matter *L2*.

Components	Tables	e_t/e	μ/μ_N	m_g/m_e	I_t	N
Nucleon crystal $\downarrow ncr_{0,0,0}^{0}$	9.1	0.0	-3.86797355	2.00	0.0	1
Nucore $\uparrow nco_{1,1,0}^{0}$	9.2	0.00	+6.69957132	1918.0000	0.0	1
Excited lambda positron $\uparrow e_{2,1,0}^{+1}$	8.15	+1.00	+0.02445099	1.000000	0.0	1
Proton $\uparrow p_{L2}^{+1}$		**+1.00**	**+2.80714679**	**1921.00000**	**0.0**	**1**
Measured values		+1.00	+ 2.79284736	1836.15267	-	-
Calculated/measured ratio		1.00	1.0051	1.0462	-	-

Stable neutron $\downarrow n_{L2}^{0}$ of the ordinary matter *L2* is made up of one proton $\uparrow p_{L2}^{+1}$ and one harmonic lambda electron $\downarrow e_{0,0,0}^{-1}$ as shown as shown in Table 9.3.2 and by the equation:

$$\downarrow n_{L2}^{0} = \uparrow p_{L2}^{+1} + \downarrow e_{0,0,0}^{-1} \qquad (9.3\text{-}2)$$

Table 9.3.2. Components and properties of the stable neutron $\downarrow n_{L2}^{0}$ of the ordinary matter *L2*.

Components	Tables	e_t/e	μ/μ_N	m_g/m_e	I_t	N
Proton $\uparrow p_{L2}^{+1}$	9.3.1	+1.00	+2.80714679	1921.0000	0.0	1
Harmonic lambda electron $\downarrow e_{0,0,0}^{-1}$	8.1	-1.00	-11.60392064	1.0000	0.0	1
Stable neutron $\downarrow n_{L2}^{0}$		**0.00**	**-8.77232286**	**1922.0000**	**0.0**	**1**

Unstable neutron $\downarrow nr_{L2}^{0}$ is a ***resonant oscillated neutron*** of the ordinary matter *L2*. It is made up of one proton $\uparrow p_{L2}^{+1}$ and one resonant oscillated electron $\downarrow er_{0,0,1}^{-1}$ as shown as shown in Table 9.3.3 and by the equation:

$$\downarrow nr_{L2}^{0} = \uparrow p_{L2}^{+1} + \downarrow er_{0,0,1}^{-1} \qquad (9.3\text{-}3)$$

Table 9.3.3. Components and properties of the unstable (resonant oscillated) neutron $\downarrow nr^0_{L2}$ of the ordinary matter *L2*.

Components	Tables	e_t/e	μ/μ_N	m_g/m_e	I_t	N
Proton $\uparrow p^{+1}_{L2}$	9.3.1	+1.00	+2.80714679	1921.0000	0.0	1
Resonant oscillated electron $\downarrow er^{-1}_{0,0,1}$	8.23	-1.00	-4.71214267	3.000000	-9.0	1
Unstable neutron $\downarrow nr^0_{L2}$		**+1.00**	**-1.90499588**	**1924.00000**	**-9.0**	**1**
Measured values		+1.00	-1.91304272	1838.68366	-	-
Calculated/measured ratio		1.00	0.9958	1.0464	-	-

It was observed that an unstable neutron decays into a proton, electron and antineutrino.

9.4 Light Atoms & Isotopes

Hydrogen atom $\downarrow^1_1 H^0_{L2}$ of the ordinary matter level *L2* is made up of one proton $\uparrow p^{+1}_{L2}$ and and one excited lambda electron $\downarrow e^{-1}_{2,1,0}$ as shown in Table 9.4.1, Fig. 9.3 and by the equation:

$$\downarrow^1_1 H^{+1}_{L2} = \uparrow p^{+1}_{L2} + \downarrow e^{-1}_{2,1,0} \tag{9.4-1}$$

Table 9.4.1. Components and properties of hydrogen atom $\downarrow^1_1 H^0_{L2}$ of the ordinary matter level *L2*.

Components	Tables	e_t/e	μ/μ_N	m_g/m_e	I_t	N
Proton $\uparrow p^{+1}_{L2}$	9.3.1	+1.00	+2.80714679	1921.0000	0.0	1
Excited lambda electron $\downarrow e^{-1}_{2,1,0}$	8.14	-1.00	-1835.658091	1.000000	0.0	1
Hydrogen atom $\downarrow^1_1 H^0_{L2}$		**0.00**	**-1832.850945**	**1922.00000**	**0.0**	**1**
Measured values		0.00	-	1838.15406	-	-
Calculated/measured ratio		1.00	-	1.0468	-	-

Deuterium $\downarrow^2_1 H^0_{L2}$ of the ordinary spacetime level *L2* is made up of one proton $\uparrow p^{+1}_{L2}$, one stable neutron $\downarrow n^0_{L2}$ and one excited lambda electron $\downarrow e^{-1}_{2,1,0}$ as shown in Table 9.4.1, Fig. 9.4.1 and by the equation:

$$\downarrow_1^2 H_{L2}^0 = \uparrow p_{L2}^{+1} + \downarrow n_{L2}^0 + \downarrow e_{2,1,0}^{-1} \qquad (9.4\text{-}1)$$

Table 9.4.1. Components and properties of the deuterium $\downarrow_1^2 H_{L2}^0$ of the ordinary matter *L2*.

Components	Tables	e_t/e	μ/μ_N	m_g/m_e	I_t	N
Proton $\uparrow p_{L2}^{+1}$	9.3.1	+1.00	+2.80714679	1921.000000	0.0	1
Stable neutron $\downarrow n_{L2}^0$	9.3.2	0.00	-8.79677385	1922.000000	0.0	1
Excited lambda electron $\downarrow e_{2,1,0}^{-1}$	8.14	-1.00	-1835.658091	1.000000	0.0	1
Deuterium $\downarrow_1^2 H_{L2}^0$		**0.00**	**-1841.647719**	**3844.000000**	**0.0**	**1**
Measured values		0.00	-	3671.485770	-	-
Calculated/measured ratio		1.00	-	1.0470	-	-

Figure 9.4.1 shows crystal structure of the deuterium nucleus. Notably, when forming nuclei containing more than one nucleon, the adjacent nucleons interlock by filling vacant vertices of their crystal structures.

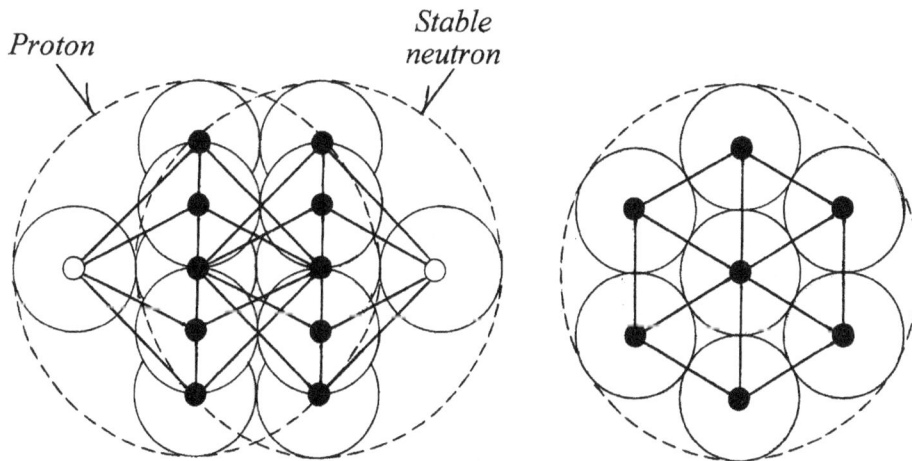

Figure 9.4.1. Crystal structure of the deuterium nucleus: front view (left) and side view (right).

Tritium $\downarrow_1^3 H_{L2}^0$ of the ordinary spacetime level *L2* is made up of one proton $\uparrow p_{L2}^{+1}$, one stable neutron $\downarrow n_{L2}^0$, one unstable neutron $\uparrow nr_{L2}^0$ and one excited lambda electron $\downarrow e_{2,1,0}^{-1}$ as shown in Table 9.4.2, Fig. 9.4.2 and by the equation:

$$\downarrow_1^3 H_{L2}^0 = \uparrow p_{L2}^{+1} + \downarrow n_{L2}^0 + \uparrow nr_{L2}^0 + \downarrow e_{2,1,0}^{-1} \qquad (9.4\text{-}2)$$

Table 9.4.2. Components and properties of the tritium $\downarrow_1^3 H_{L2}^0$

of the ordinary matter *L2*.

Components	Tables	e_t/e	μ/μ_N	m_g/m_e	I_t	N
Proton $\uparrow p_{L2}^{+1}$	9.3.1	+1.00	+2.80714679	1921.000000	0.0	1
Stable neutron $\downarrow n_{L2}^0$	9.3.2	0.00	-8.79677385	1922.000000	0.0	1
Unstable neutron $\downarrow nr_{L2}^0$	9.3.3	0.00	-1.90499588	1924.000000	-9.0	1
Excited lambda electron $\downarrow e_{2,1,0}^{-1}$	8.14	-1.00	-1835.658091	1.000000	0.0	1
Deuterium $\downarrow_1^2 H_{L2}^0$		**0.00**	**-1843.552720**	**5768.000000**	**-9.0**	**1**
Measured values		-	-	5497.925226	-	-
Calculated/measured ratio		-	-	1.0491	-	-

Note: Tritium is unstable, because it contains the unstable neutron $\downarrow nr_{L2}^0$,

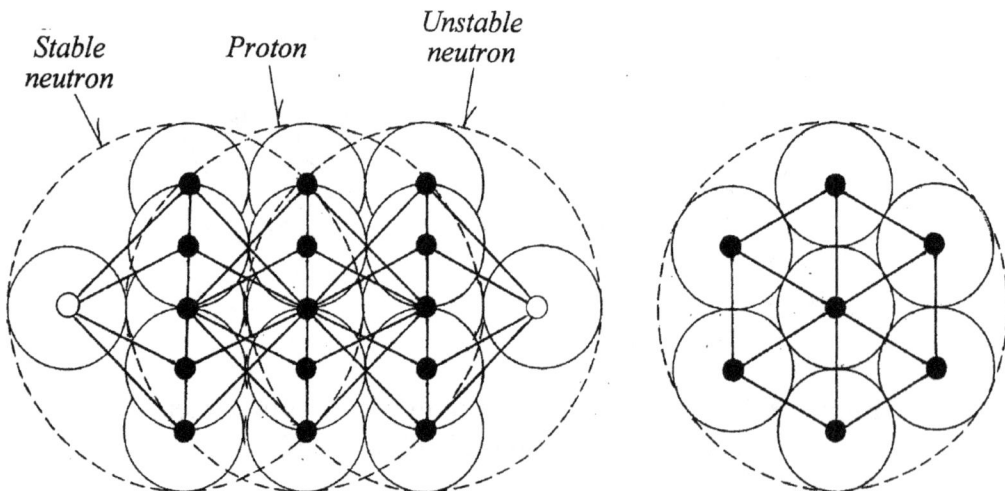

Figure 9.4.2. Crystal structure of the tritium nucleus:
front view (left) and side view (right).

Helium-3 $\downarrow {}_2^3 He_{L2}^0$ of the ordinary spacetime level *L2* is made up of two protons p_{L2}^{+1} with opposite signs of their magnetic moments, one stable neutron $\downarrow n_{L2}^0$ and two excited lambda electrons $e_{2,1,0}^{-1}$ with opposite signs of their magnetic moments as shown in Table 9.4.3, Fig. 9.4.3 and by the equation:

$$\downarrow {}_2^3 He_{L2}^0 = 2 \updownarrow p_{L2}^{+1} + \downarrow n_{L2}^0 + 2 \updownarrow e_{2,1,0}^{-1} \qquad (9.4\text{-}3)$$

Table 9.4.3. Components and properties of the helium-3 $\downarrow {}_2^3 He_{L2}^0$
of the ordinary matter *L2*.

Components	Table	e_t/e	μ/μ_N	m_g/m_e	I_t	N
Protons $\updownarrow p_{L2}^{+1}$	9.3.1	+1.00	± 2.80714679	1921.000000	0.0	2
Stable neutron $\downarrow n_{L2}^0$	9.3.2	0.00	-8.79677385	1922.000000	0.0	1
Excited lambda electrons $\updownarrow e_{2,1,0}^{-1}$	8.14	-1.00	± 1835.658091	1.000000	0.0	2
Helium-3 $\downarrow {}_2^3 He_{L2}^0$		**0.00**	**- 8.772323**	**5766.000000**	**0.0**	**1**
Measured values		-	-	5497.888768	-	-
Calculated/measured ratio		-	-	1.0488	-	-

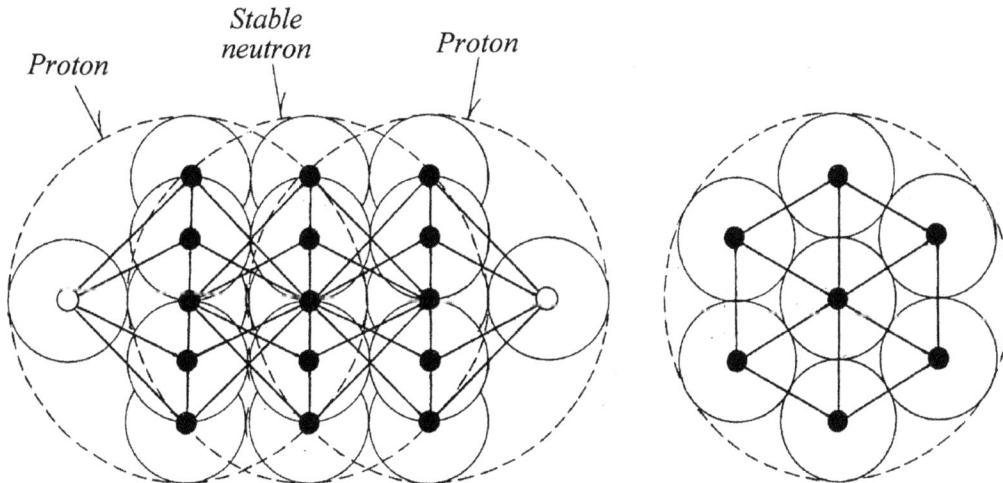

Figure 9.4.3. Crystal structure of the helium-3 nucleus:
front view (left) and side view (right).

Helium-4 $^4_2He^0_{L2}$ of the spacetime level L is made up of two protons p^{+1}_{L2}, two stable neutrons n^0_{L2} and two excited lambda electrons $e^{-1}_{2,1,0}$, with particles in all three pairs having opposite signs of their magnetic moments as shown in Table 9.4.4, Fig. 9.4.4 and by the equation:

$$\downarrow ^4_2He^0_{L2} = 2 \updownarrow p^{+1}_{L2} + 2 \updownarrow n^0_{L2} + 2 \updownarrow e^{-1}_{2,1,0} \tag{9.4-4}$$

Table 9.4.4. Components and properties of the helium-4 $^4_2He^0_{L2}$
of the ordinary matter *L2*.

Components	Table	e_t/e	μ/μ_N	m_g/m_e	I_t	N
Protons $\updownarrow p^{+1}_{L2}$	9.3.1	+1.00	± 2.80714679	1921.000000	0.0	2
Stable neutron $\downarrow n^0_{L2}$	9.3.2	0.00	± 8.79677385	1922.000000	0.0	2
Excited lambda electrons $\updownarrow e^{-1}_{2,1,0}$	8.14	-1.00	± 1835.658091	1.000000	0.0	2
Helium-4 $^4_2He^0_{L2}$		**0.00**	**0.00**	**7688.000000**	**0.0**	**1**
Measured values		-	-	2296.303079	-	-
Calculated/measured ratio		-	-	1.0537	-	-

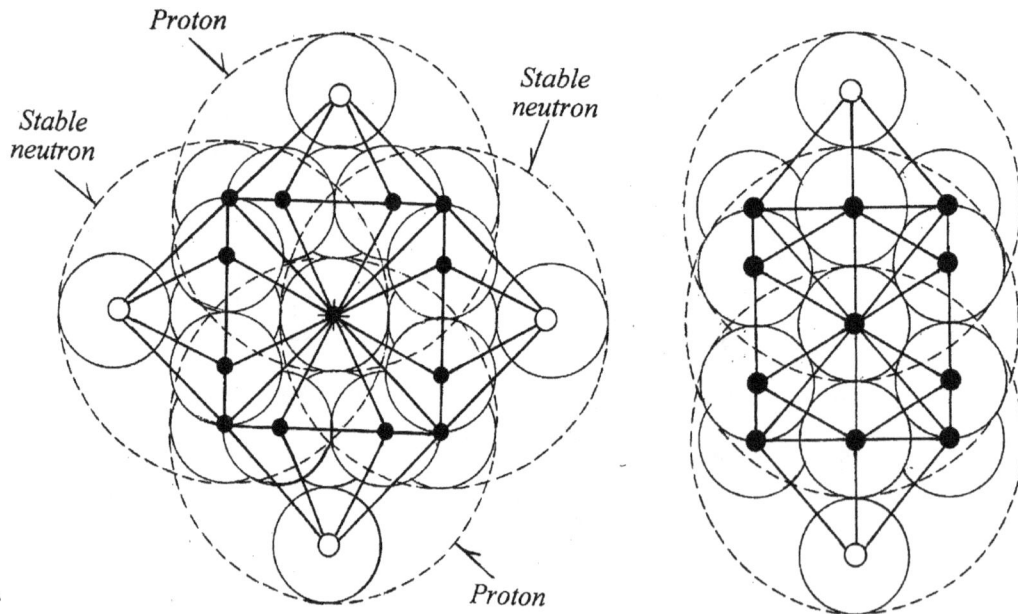

Figure 9.4.4 Crystal structure of the helium-4 nucleus: front view (left)
and side view (right).

9.5 Effect of Spacetime Levels

Both the compositions and properties of protons are greatly dependent on their spacetime levels L. Table 9.5.1 shows some physical properties of stable protons of the spacetime levels $L1$, $L2$ and $L3$.

- Proton mass of the spacetime level $L1$ is 113 times less than the proton mass of the spacetime level $L2$
- Proton mass of the spacetime level $L3$ is 136.9 times greater than the proton mass of the spacetime level $L2$.

Table 9.5.1. Composition and properties of the stable protons $\uparrow p_L^{+1}$ of several spacetime levels L.

Spacetime level	Proton	μ / μ_N	m_g / m_e	Mass ratio
$L1$	$\uparrow p_{L1}^{+1}$	+ 3.58152247	17.0	1/113
$L2$ (ordinary matter)	$\uparrow p_{L2}^{+1}$	+ 2.80714679	1921.0	1.00
$L3$	$\uparrow p_{L3}^{+1}$	+ 2.82946437	262769.0	136.9

Table 9.5.2. Relative parameters of the hydrogen atoms $\downarrow H_L^0$ in the ground state $n = 1$ of several spacetime levels L.

Spacetime levels L	Hydrogen atoms	Electron orbital radius ratio	Orbital magnetic moment ratio	Hydrogen atom mass ratio
$L1$	$\downarrow H_{L1}^0$	1/137.25	1/11.7	1/106.8
$L2$ (ordinary matter)	$\downarrow H_{L2}^0$	1.0	1.0	1.0
$L3$	$\downarrow H_{L3}^0$	137.0	11.7	136.7

Both the compositions and properties of hydrogen atoms are greatly dependent on their spacetime levels L. Table 9.5.2 shows some physical properties of hydrogen atoms aof the spacetime levels $L1$, $L2$ and $L3$.

- Orbital radius of an atomic electron located at the ground state of a hydrogen atom of the spacetime level $L1$ is 137.25 times smaller than a respective orbital radius of an electron of a hydrogen atom of the spacetime level $L2$ (ordinary matter)
- Orbital radius of an atomic electron located at the ground state of a hydrogen atom of the spacetime level $L3$ is 137 times larger than a respective orbital radius of a hydrogen atom of the spacetime level $L2$ (ordinary matter).

- Orbital magnetic moment of an atomic electron located at the ground state of a hydrogen atom of the spacetime level *L1* is 11.7 times smaller than a respective orbital radius of an electron of a hydrogen atom of the spacetime level *L2* (ordinary matter)
- Orbital magnetic moment of an atomic electron located at the ground state of a hydrogen atom of the spacetime level *L3* is 11.7 times larger than a respective orbital radius of a hydrogen atom of the spacetime level *L2* (ordinary matter)
- Mass of a hydrogen atom of the spacetime level *L1* is 106.8 times smaller than a respective mass of a hydrogen atom of the spacetime level *L2* (ordinary matter)
- Mass of a hydrogen atom of the spacetime level *L3* is 136.7 times greater than a respective mass of a hydrogen atom of the spacetime level *L2* (ordinary matter).

10. MACRO-TORYCES & MACRO-TRONS

In previous chapters we described a role of the ***micro-toryces*** in the formation of matter particles and atoms of the micro-world. In the macro-world, the assemblies of atoms contained in each body form the macro-spacetimes called the ***macro-toryces*** that become integral parts of each body.

10.1 Basic Structure & Parameters of Macro-Toryces

Figure 10.1 shows a macro-toryx associated with the central body A and encompassing the satellite body B.

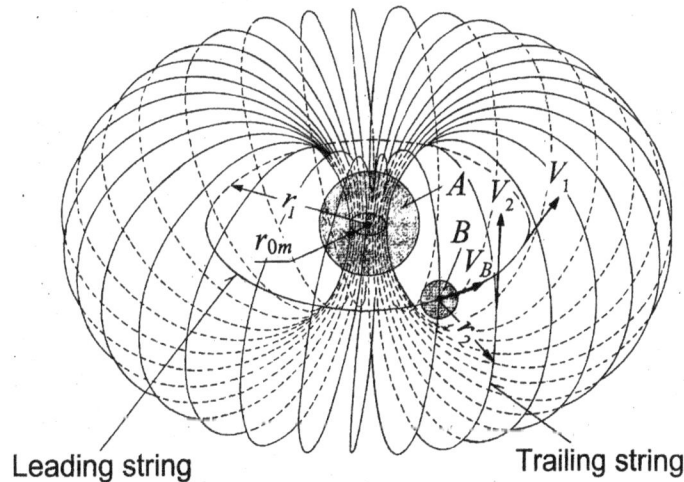

Figure 10.1. Central body A with a macro-toryx encompassing the satellite body B.

Similarly to the micro-toryx (Figs. 1.1.1 – 1.1.3), the macro-toryx contains two strings, a ***leading string*** and a ***trailing string***. Leading string is double-circular; it is moving with the velocity V_1 along a circle with the radius r_1. Trailing string is double-toroidal with the radius r_2 in which each branch propagates along its toroidal spiral path with the spiral velocity V_2 that has two components, the translational velocity V_{2t} and the rotational velocity V_{2r}.

10.2 Macro-Toryx Spacetime Postulates

Exhibit 10.2.1 shows the spacetime postulates expressed in absolute spacetime units. These postulates are the same as those used for micro-toryces shown in Exhibit 1.3.

Exhibit 10.2.1. Macro-toryx spacetime postulates in absolute units.

- The length of one winding of trailing string L_2 is equal to the length of one winding of leading string L_1:

$$L_2 = L_1 = 2\pi r_1 \qquad (10.2\text{-}1)$$

- The macro-toryx eye radius r_{0m} is constant:

$$r_{0m} = r_1 - r_2 = const. \qquad (10.2\text{-}2)$$

- The spiral velocity of trailing string V_2 is constant at each point of its spiral path:

$$V_2 = \sqrt{V_{2t}^2 + V_{2r}^2} = c = const. \qquad (10.2\text{-}3)$$

Similarly to the micro-toryx, the spacetime postulates of the macro-toryces can also be simplified (Exhibit 10.2.2) by expressing the macro-toryx spacetime parameters in relative units in respect to the constant macro-toryx parameters: the eye radius r_{0m} and the velocity of light c.

Exhibit 10.2.2. Toryx spacetime postulates in relative units.

- The relative length of one winding of trailing string l_2 is equal to the relative length of one winding of leading string l_1:

$$l_2 = l_1 \qquad (10.2\text{-}4)$$

- The macro-toryx relative eye radius b_{0m} is equal to 1:

$$b_{0m} = b_1 - b_2 = 1 \qquad (10.2\text{-}5)$$

- The relative spiral velocity of trailing string β_2 is equal to 1 at each point of its spiral path:

$$\beta_2 = \sqrt{\beta_{2t}^2 + \beta_{2r}^2} = 1 \qquad (10.2\text{-}6)$$

10.3 Classification of Macro-Toryces & Macro-Trons

Similarly to the classification of the micro-toryces shown in Chapter 3, the macro-toryces are also divided into four groups:

- Real negative macro-toryces
- Real positive macro-toryces
- Imaginary positive macro-toryces
- Imaginary negative macro-toryces.

Similarly to the classification of the micro-trons shown in Section 7.3, the macro-trons are also divided into four groups:

- Macro-electrons
- Macro-positrons
- Macro-ethertrons
- Macro-singulatrons.

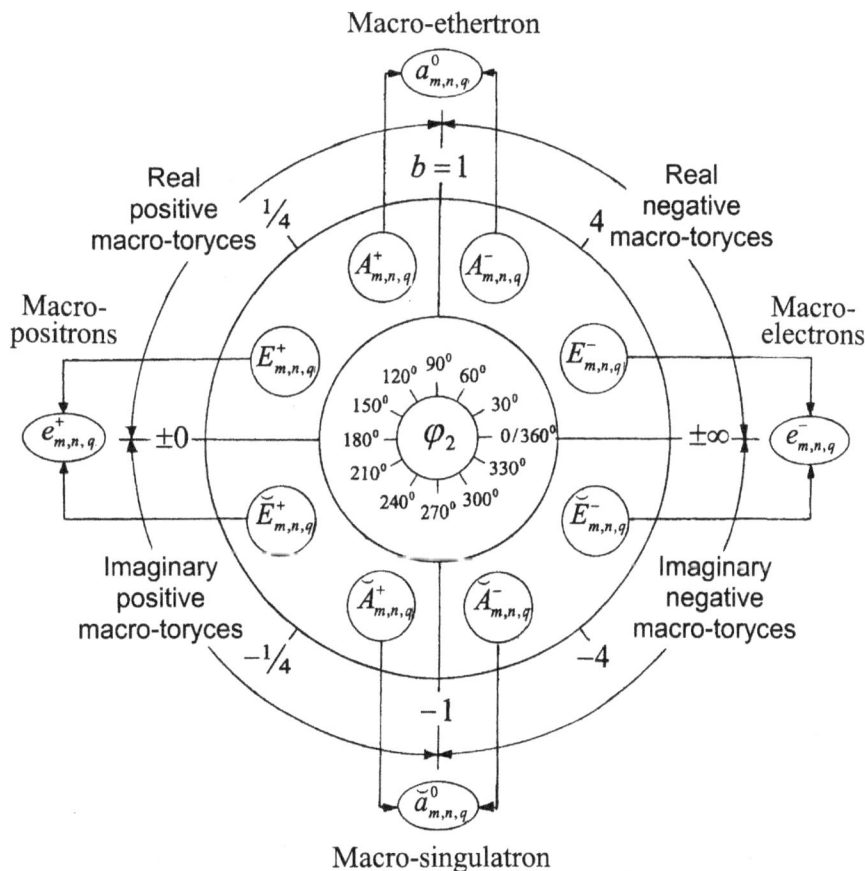

Figure 10.3. Formation of four macro-trons from polarized macro-toryces.

10.4 Spacetime Law of Planetary Motion

In classical mechanics, the law of planetary motion of the macro-world is derived by equating two forces (Fig. 10.4):

- The attraction gravitational force F_g between the central body A with the mass m_A and the body B with the mass m_B separated by the distance r_1

- The centrifugal forces F_c applied to the body B with the mass m_B orbiting the central body with the orbital velocity V_1.

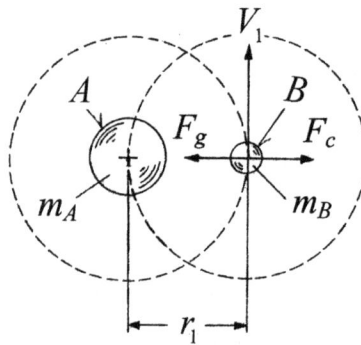

Figure 10.4. A planetary system in which the body B with the mass m_B orbits the central body A with the mass m_A.

Based on Newton's universal law of gravitation and an equation for the centrifugal force, we obtain that at the equilibrium state:

$$F_g = F_c = \frac{m_A m_B G}{r_1^2} = \frac{m_B V_1^2}{r_1} \tag{10.4-1}$$

Consequently, we obtain from Eq. (10.4-1) the classical equation for the law of planetary motion of the macro-world:

$$V_1 = \sqrt{\frac{m_A G}{r_1}} \tag{10.4-2}$$

Let us define the relative distance b_1 between the celestial bodies A and B in respect to the eye radius r_{0m} of the macro-toryx associated with the body A by the equation:

$$b_1 = \frac{r_1}{r_{0m}} \tag{10.4-3}$$

The classical (10.4-2) for the law of planetary motion of the macro-world is applicable to a case when the distance r_1 is significantly greater than the eye radius r_{0m} of the macro-toryx associated with the body A, so $r_1 \gg r_{0m}$. To extend the application of this equation to much shorter distances, let us apply Eq. (1.7-1) for the relative orbital velocity β_1 of the body B derived for the macro-toryx:

$$\beta_1 = \frac{V_1}{c} = \frac{\sqrt{2b_1 - 1}}{b_1} \qquad (10.4\text{-}4)$$

Thus, Eq. (10.4-4) describes the *spacetime law of planetary motion* of the macro-world applicable to any distances r_1. This equation reduces to a classical equation when $r_1 \gg r_{0m}$:

$$\beta_1 = \frac{V_1}{c} = \sqrt{\frac{2r_{0m}}{r_1}} \qquad (10.4\text{-}5)$$

From Eqs. (10.4-2) and (10.4-5), we can find that the eye radius r_{0m} of the macro-toryx associated with the body A is equal to:

$$r_{om} = \frac{m_A G}{2c^2} \qquad (10.4\text{-}6)$$

Kepler's third Law of planetary motion - From Eqs. (10.4-3) and (10.4-4) and by considering that $V_1 = 2\pi r_1 / T_1$ we may express the spacetime law of planetary motion in the form:

$$r_1^3 = k T_1^2 \left(1 - \frac{r_{0m}}{2r_1} \right) \qquad (10.4\text{-}7)$$

where the constant k is equal to:

$$k = \frac{r_{0m} c^2}{2\pi^2} = \frac{m_A G}{4\pi^2} \qquad (10.4\text{-}8)$$

Thus, for the case when the radius of this string r_1 is much greater than the macro-toryx eye radius r_{0m}, Eq. (10.4-7) reduces to the Kepler's third law of planetary motion:

$$r_1^3 = k T_1^2 \qquad (10.4\text{-}9)$$

10.5 Interaction Between Celestial Bodies

We proposed that the macro-trons associated with celestial bodies are responsible for the interactions between these bodies. Figure 10.5 shows two adjacent celestial bodies A and B and the respective macro-toryces A and B associated with these bodies.

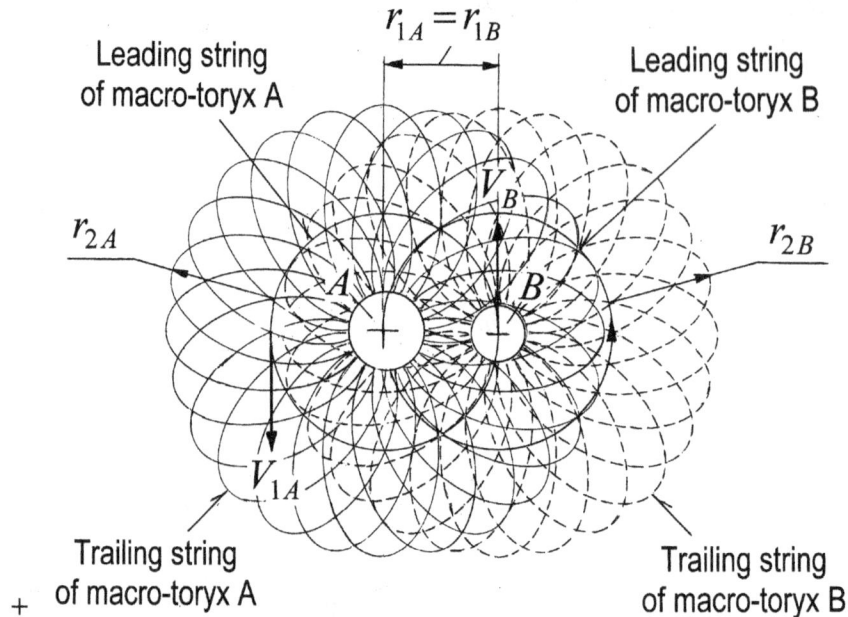

Figure 10.5. Two adjacent celestial bodies *A* and *B* with macro-toryces A and B associated with these bodies.

Let the macro-toryx A to be associated with the body *A* having the mass m_A. The macro-toryx circular leading string with the radius r_{1A} propagates around a center of the body *A* with the orbital velocity V_{1A}, while its toroidal trailing string with the radius r_{2B} propagates synchronously with the leading string. Located at the distance r_{1A} from the center of the body *A* is a center of the body *B* having the mass m_B. The body *B* moves around the center of the body *A* along a circular path with the radius $r_{1B} = r_{1A}$ with the orbital velocity V_B.

Exhibit 10.5. Spacetime acceleration of a body.

Spacetime acceleration a_{bs} of the body *B* in respect to a center of the body *A* is equal to:

$$a_{bs} = \frac{V_{1A}^2 - V_B^2}{r_{1A}} = \frac{V_{1A}^2(1 - \gamma_V^2)}{r_{1A}} \tag{10.5-1}$$

where γ_V is the velocity ratio that is equal to:

$$\gamma_V = \frac{V_B}{V_1} \tag{10.5-2}$$

The behavior of the body B depends on a relationship between its orbital velocity V_B and the spiral velocity of leading string V_{1A} of the macro-toryx A associated with the body A. When these velocities are not exactly the same, it means the body B does not follow the spacetime law of planetary motion described by Eq. (10.4-4). Consequently, the body B will move in a radial direction either towards to or away from the body A with the *spacetime acceleration* a_{bs} described by the proposed equation shown in Exhibit 10.5.

Considering Eqs. (10.4-3), (10.4-4), (10.4-6) and (10.5-1), the spacetime acceleration a_{bs} of the body B in respect to the body A separated by the distance b_{1A} is equal to:

$$a_{bs} = \frac{m_A G}{r_{1A}^2} \frac{2b_{1A} - 1}{2b_{1A}} (1 - \gamma_V^2) \tag{10.5-3}$$

For a particular case when $b_{1A} >> 1$ and $\gamma_V = 0$, Eq. (10.5-3) reduces to the form:

$$a_b = \frac{m_A G}{r_{1A}^2} \tag{10.5-4}$$

Based on Eqs. (10.5-1) - (10.5-4), we may conclude:

- When the orbital velocity V_B of the body B around a center of the body A is the same as the velocity of propagation of leading string V_{1A} of the macro-toryx associated with the body A ($\gamma_V = 1$), the body B will orbit the body A according to the spacetime law of planetary motion and, consequently, the velocity V_B and its distance to the body B will remain unchanged.
- When the velocity V_B of the body B around the body A is different than the velocity of propagation of leading string V_1 of the macro-toryx associated with the body A, the body B accelerates either towards to or away from the body A.

10.6 Spacetime Law of Gravitation

Consider a case when the satellite body B with the mass m_B does not comply with the spacetime law of planetary motion and moves freely either towards to or away from the body A with the acceleration a_b. Let the dimensions of the body B to be negligibly small in comparison with its distance from the body A. In that case, the body B will not experience any force applied to it until after its free motion is affected by either internal or external causes. In that moment a holding force will be applied to the body B.

This holding force is called the *spacetime gravitational force* F_{gs}, and it is equal to the product of the body mass m_B and its acceleration a_{bs} as described by the equation:

$$F_{gs} = m_B a_{bs} \tag{10.6-1}$$

Similarly to Eq. (7.1-5) applied to a toryx, when $Q_q = 1$, the inertial mass of the body B is equal to:

$$m_{Bi} = m_B \frac{2(b_{1A} - 1)}{2b_{1A} - 1} \qquad (10.6\text{-}2)$$

Consequently, we obtain from Eqs. (10.5-3), (10.6-1) and (10.6-2) that the spacetime gravitational force F_{gs} is equal to:

$$F_{gs} = \frac{m_A m_B G}{r_{1A}^2} \frac{(b_{1A} - 1)}{b_{1A}} (1 - \gamma_V^2) \qquad (10.6\text{-}3)$$

We called Eq. (10.6-3) the *spacetime law of gravitation*. For a particular case when $b_1 \gg 1$ and $\gamma_V^2 \ll 1$, Eq. (10.6-3) reduces to a familiar equation expressing the Newton's universal law of gravitation:

$$F_{gN} = \frac{m_A m_B G}{r_{1A}^2} \qquad (10.6\text{-}4)$$

Based on Eqs. (10.5-1) - (10.5-3), (10.6-3) and Fig. 10.6, we may conclude:

- When the velocity V_B of the satellite body B around the central body A is the same as the velocity of propagation of the leading string V_{1A} of the macro-toryx associated with the central body A ($\gamma_V = 1$), the body B will orbit the body A according to the spacetime law of planetary motion and, consequently, the velocity V_B and its distance to the satellite body B will remain unchanged.

- When the velocity V_B of the satellite body B around the central body A is different than the velocity of propagation of the leading string V_{1A} of the macro-toryx associated with the central body A, the satellite body B accelerates either towards to or away from the body A.

- If the radius of the satellite body B is negligibly small than its distance to the central body A then the "gravitational force" will be "felt" by the satellite body B only after its acceleration towards or away from the central body A will be obstructed by either internal or external means.

Figure 10.6 shows the ratio of the spacetime to Newton's gravitational force F_{gs} / F_{gN} defined by Eqs. (10.6-3) and (10.6-4) as a function of the relative distance b_1 between the bodies A and B when velocity of the body B is equal to zero ($\gamma_V = 0$).

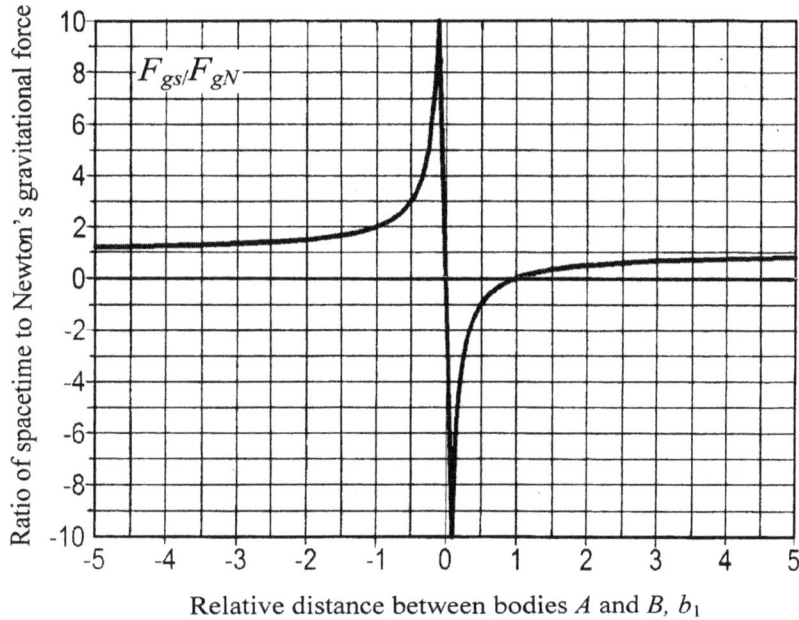

Figure 10.6. Ratio of the spacetime to Newton's gravitational force F_{gs} / F_{gN} as a function of the relative distance b_1 between bodies A and B.

10.7 Quantum States of Macro-Toryces of Celestial Bodies

Quantum states of macro-toryces of celestial bodies are described by the same quantization equations as the quantization equations of micro-toryces shown in Table 6.2. The quantization parameter z of a macro-toryx is expressed in Exhibit 10.7.

Exhibit 10.7. The quantization parameter z of a macro-toryx.

The quantization parameter z of a macro-toryx is expressed by the equation:

$$z = 2(k_b n \Lambda)^2 \qquad (10.7\text{-}1)$$

where k_b is the constant celestial body parameter.

Tables 10.7.1 – 10.7.5 show parameters of planets and moons in our solar system, and also their possible past, current and projected future status. Table 10.7.6 shows the values of the celestial body parameters k_b, the body masses m_A, the equatorial body radii r_b and the mass eye radii r_{0m} for the largest bodies of our solar system. Notably, the celestial body parameters k_b is probably dependent on both density and distribution of mass inside the body.

Table 10.7.1. Parameters and status of planets in our solar system.

n	Planets	Measured distance to Sun r_{1m}, m	Relative measured dist. to Sun b_{1m}	Relative calculated dist. to Sun b_{1c}	Ratio b_{1m}/b_{1c}	Status of Sun's planets
1	Sun1	6.2378×10^{09}	8.4458×10^6	8.4458×10^6	1.0000	Former planets swallowed by the Sun
2	Sun 2	2.4951×10^{10}	3.3784×10^7	3.3784×10^7	1.0000	
3	**Mercury**	5.7909×10^{10}	7.8408×10^7	7.6014×10^7	**1.0315**	**Currently existing**
4	**Venus**	1.0821×10^{11}	1.4651×10^8	1.3514×10^8	**1.0842**	
5	**Earth**	1.4960×10^{11}	2.0255×10^8	2.1115×10^8	**0.9593**	
6	**Mars**	2.2794×10^{11}	3.0863×10^8	3.0406×10^8	**1.0150**	
7	Hungaria	3.0567×10^{11}	4.1387×10^8	4.1387×10^8	1.0000	Future planets that could be formed from asteroid belts
8	Phocaea	3.9923×10^{11}	5.4055×10^8	5.4055×10^8	1.0000	
9	Cybele	5.0527×10^{11}	6.8412×10^8	6.8412×10^8	1.0000	
10	Hilda	6.2382×10^{11}	8.4464×10^8	8.4464×10^8	1.0000	
11	**Jupiter**	7.7833×10^{11}	1.0538×10^9	1.0220×10^9	**1.0312**	**Currently existing**
12	Io	8.9826×10^{11}	1.2162×10^9	1.2162×10^9	1.0000	Former planets captured by the Jupiter and became its moons
13	Europe	1.0542×10^{12}	1.4273×10^9	1.4273×10^9	1.0000	
14	Ganymede	1.2227×10^{12}	1.6555×10^9	1.6555×10^9	1.0000	
15	**Saturn**	1.4270×10^{12}	1.9321×10^9	1.9004×10^9	**1.0167**	**Currently existing**
16	Tethys	1.5970×10^{12}	1.1622×10^9	1.1622×10^9	1.0000	Former planets captured by the Saturn and became its moons
17	Dione	1.1029×10^{12}	2.4410×10^9	2.4410×10^9	1.0000	
18	Rhea	2.0210×10^{12}	2.7364×10^9	2.7364×10^9	1.0000	
19	Titan	2.2518×10^{12}	3.0489×10^9	3.0489×10^9	1.0000	
20	Iapetus	2.4951×10^{12}	3.3783×10^9	3.3783×10^9	1.0000	
21	**Uranus**	2.8696×10^{12}	3.8853×10^9	3.7247×10^9	**1.0431**	**Currently existing**
22	Miranda	3.0190×10^{12}	4.0877×10^9	4.0877×10^9	1.0000	Former planets captured by the Uranus and became its moons
23	Ariel	3.2998×10^{12}	4.4678×10^9	4.4678×10^9	1.0000	
24	Umbriel	3.5929×10^{12}	4.8647×10^9	4.8647×10^9	1.0000	
25	Titania	3.8988×10^{12}	5.2789×10^9	5.2789×10^9	1.0000	
26	Oberon	4.2170×10^{12}	5.7097×10^9	5.7097×10^9	1.0000	
27	**Neptune**	4.4966×10^{11}	6.0883×10^9	6.0883×10^9	**0.9888**	**Currently existing**
28	Proteus	4.8904×10^{12}	6.6214×10^9	6.6214×10^9	1.0000	Former planets captured by the Neptune and became its moons
29	Triton	5.2460×10^{12}	7.1029×10^9	7.1029×10^9	1.0000	

Table 10.7.2. Parameters and status of the Jupiter's moons.

n	Jupiter's moons	Measured distance to Jupiter r_{1m}, m	Relative measured distance to Jupiter b_{1m}	Relative calculated distance to Jupiter b_{1c}	Ratio b_{1m}/b_{1c}	Status of Jupiter's moons
1	Jupiter 1	2.7115×10^7	3.8439×10^7	3.8439×10^7	1.0000	Former moons swallowed by the Jupiter
2	Jupiter 2	1.0846×10^8	1.5376×10^8	1.5376×10^8	1.0000	
3	Io	2.6200×10^8	3.7143×10^8	3.4595×10^8	1.0737	Currently existing moons
4	Europa	4.1690×10^8	5.9103×10^8	6.1502×10^8	0.9610	
5	Ganymede	6.6490×10^8	9.4261×10^8	9.6097×10^8	0.9809	
7	Callisto	1.1701×10^9	1.6588×10^9	1.8835×10^9	0.8807	

Table 10.7.3. Parameters and status of the Saturn's moons.

n	Saturn's moons	Measured distance to Saturn r_{1m}, m	Relative measured distance to Saturn b_{1m}	Relative calculated distance to Saturn b_{1c}	Ratio b_{1m}/b_{1c}	Status of Saturn's moons
1	Saturn 1	2.3974×10^7	1.1355×10^8	1.1355×10^8	1.0000	Former moons swallowed by the Saturn
2	Saturn 2	9.5890×10^7	4.5421×10^8	4.5421×10^8	1.0000	
3	Tethys	2.9462×10^8	1.3955×10^9	1.0220×10^9	1.3655	Currently existing moons
4	Dione	3.7740×10^8	1.7876×10^9	1.8168×10^9	0.9839	
5	Rhea	5.2711×10^8	2.4967×10^9	2.8388×10^9	0.8795	
7	Titan	1.2219×10^9	5.7877×10^9	5.5641×10^9	1.0402	
12	Iapetus	3.5608×10^9	1.6866×10^{10}	1.6352×10^{10}	1.0315	

Table 10.7.4. Parameters and status of the Uranus' moons.

N	Uranus' moons	Measured distance to Uranus r_{1m}, m	Relative measured distance to Uranus b_{1m}	Relative calculated distance to Uranus b_{1c}	Ratio b_{1m}/b_{1c}	Status of Uranus' moons
1	Uranus 1	1.2111×10^7	3.7538×10^8	3.7538×10^8	1.0000	Former moons swallowed by the Uranus
2	Uranus 2	4.8446×10^7	1.5015×10^9	1.5015×10^9	1.0000	
3	Miranda	1.2978×10^8	4.0225×10^9	3.3784×10^9	1.1907	Currently existing moons
4	Ariel	1.9102×10^8	5.9207×10^9	6.0061×10^9	0.9858	
5	Umbriel	2.6630×10^8	8.2540×10^9	9.3845×10^9	0.8795	
6	Titania	4.3591×10^8	1.3511×10^{10}	1.3514×10^{10}	0.9998	
7	Oberon	5.8352×10^8	1.8086×10^{10}	1.8394×10^{10}	0.9833	

Table 10.7.5. Parameters and status of the Neptun's moons.

n	Neptune's moons	Measured distance to Neptune r_{1m}, m	Relative measured distance to Neptune b_{1m}	Relative calculated distance to Neptune b_{1c}	Ratio b_{1m}/b_{1c}	Status of Neptune's moons
1	Neptune 1	1.4352×10^7	3.7538×10^8	3.7538×10^8	1.0000	Former moons swallowed by the Neptune
2	Neptune 2	4.8446×10^7	1.5015×10^9	1.5015×10^9	1.0000	
3	Proteus	1.1765×10^8	3.0772×10^9	3.3784×10^9	0.9108	Currently existing moons
5	Triton	3.5476×10^8	9.2791×10^9	9.3845×10^9	0.9888	

Table 10.7.6. Parameters of largest celestial bodies of our solar system.

Celestial body	k_b	m_A, kg	r_b, m	r_{0m}, m
Sun	15	1.98910×10^{30}	6.9550×10^8	738.566556
Jupiter	32	1.89973×10^{27}	7.1492×10^7	0.705383
Saturn	55	5.68598×10^{26}	6.0268×10^7	0.211124
Uranus	100	8.68910×10^{25}	2.5559×10^7	0.032263
Neptune	100	1.02966×10^{26}	2.4764×10^7	0.038232

Macro-tons are the radiation particles of the macro-world emitted by their paternal macro-trons. They are composed of reality-polarized and charge-polarized matched macro-helyces. The macro-tons are emitted after an orbit of either a planet or a satellite collapses to a lower quantum state. The macro-trons are probably analogous to what we called nowadays the *gravitons*.

10.8 Spacetime Law of Star Motion

Consider a star moving along a circular path with the radius r_s around a galaxy center with the orbital velocity V_s as shown in Fig. 10.8.

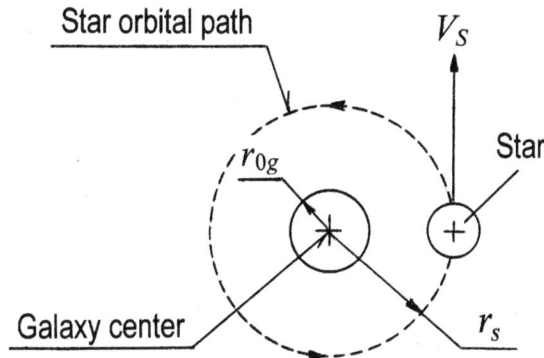

Figure 10.8. A star moving along a circular path with the radius r_s around a galaxy center with the orbital velocity V_s.

In that case, the eye radius of the macro-toryx r_{0g} associated with this star is defined by making two assumptions shown in Exhibit 10.8.

Exhibit 10.8. The eye radius of a macro-toryx r_{0g} associated with a star.

- The eye radius of a macro-toryx r_{0g} associated with a star is directly-proportional to the galaxy mass m_G located within a sphere with the radius equal to the star orbital radius r_s:

$$r_{0g} = \frac{m_G G}{2c^2} \qquad (10.8\text{-}1)$$

- When the star orbital radius r_s is much greater than the radius of galaxy's black hole, the ratio s of the galaxy mass m_G to the star orbital radius r_s is constant:

$$\frac{m_G}{r_s} = s = const. \qquad (10.8\text{-}2)$$

For the case when the star orbital radius r_s is much greater than the radius of the galaxy's black hole, Eq. (10.4-5) reduces to the form representing the spacetime law of star motion in relative units:

$$\beta_s = \frac{V_s}{c} = \sqrt{\frac{2}{b_s}} \qquad (10.8\text{-}3)$$

where from Eqs. (10.5-1) and (10.5-2) the relative star orbital radius b_s is equal to:

$$b_s = \frac{r_s}{r_{0g}} = \frac{2c^2}{sG} = const. \qquad (10.8\text{-}4)$$

Consequently, the spacetime law of star motion expressed by Eq. (10.8-3) reduces the form representing the spacetime law of star motion in absolute units::

$$V_s = \sqrt{\frac{m_G G}{r_s}} = \sqrt{sG} = const. \qquad (10.8\text{-}5)$$

Thus, for the case when the star orbital radius r_s is much greater than the radius of the galaxy's black hole, the star velocity V_s is constant and does not depend on the star orbital radius r_s.

10.9 Outverted & Inverted Stars

Depending on the relationship between the outer radius r_b of a star body and the eye radius r_{0m} of a macro-toryx associated with a star, the stars can be divided into three kinds: *outverted stars*, *inverted stars* and *imaginary stars* as shown in Figure 10.9.

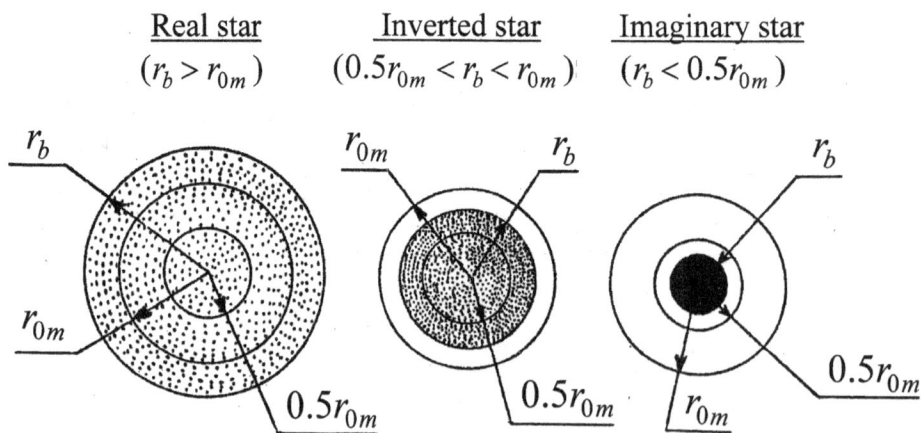

Figure 10.9. Three kinds of stars.

Outverted stars $(r_b > r_{0m})$ – In the outverted stars, the outer radius r_b of the star body is greater than the eye radius r_{0m} of a macro-toryx associated with this star. Our Sun is a typical example of the outverted star.

Inverted stars $(0.5r_b < r_b < r_{0m})$ – In the inverted stars, the outer radius r_b of a star body is greater than one half of the eye radius r_{0m}, but less than the eye radius r_{0m} of macro-toryx associated with this star.

Imaginary stars $(r_b < 0.5r_b)$ – In the imaginary stars, the outer radius r_b of a star body is less than one half of the eye radius r_{0m} of a macro-toryx associated with this star.

<u>*Notes*</u>

PHYSICAL & SPACETIME CONSTANTS

Physical Constants

Names		Values
Elementary charge	e	$1.602\ 176\ 565 \times 10^{-19}$ C
Electric constant	ε_0	$8.854\ 187\ 817 \times 10^{-12}$ C^2/N/m^2
Electron mass	m_e	$9.109\ 382\ 91 \times 10^{-31}$ kg
Proton relative mass	m_p / m_e	$1836.152\ 671\ 95$
Neutron relative mass	m_n / m_e	$1838.683\ 660\ 08$
Muon relative mass	m_μ / m_e	$206.768\ 284\ 26$
Tau relative mass	m_τ / m_e	$3477.151\ 011\ 54$
Newtonian constant of gravitation	G	$6.674\ 280 \times 10^{-11}$ m^3/kg/s^2
Planck constant	h	$6.626\ 069\ 605 \times 10^{-34}$ J s
Bohr magneton	μ_B	$9.274\ 009\ 68 \times 10^{-24}$ J/T
Muon magneton	μ_μ	$4.485\ 218\ 58 \times 10^{-26}$ J/T
Tau magneton	μ_τ	$2.666\ 876\ 44 \times 10^{-24}$ J/T
Electron magnetic moment to Bohr magneton ratio	μ_e / μ_B	$-1.001\ 159\ 652\ 180\ 76$
Muon magnetic moment to muon magneton ratio	μ_μ / μ_μ	$-1.001\ 165\ 923$
Tau magnetic moment to tau magneton ratio	μ_τ / μ_τ	$-0.987\ 629\ 486$
Nuclear magneton	μ_N	$5.050\ 783\ 54 \times 10^{-27}$ J/T
Proton magnetic moment to nuclear magneton ratio	μ_p / μ_N	$2.792\ 847\ 356$
Neutron magnetic moment to nuclear magneton ratio	μ_n / μ_N	$-1.913\ 042\ 72$

Physical & Spacetime Constants

Spacetime Constants

Names		Values
Speed of light in vacuum	c	$2.997\ 924\ 58 \times 10^{08}$ m/s
Classical electron radius	r_e	$2.817\ 940\ 327 \times 10^{-15}$ m
Micro-toryx eye radius	r_0	$1.408\ 970\ 164 \times 10^{-15}$ m
Toryx basic frequency	f_0	$3.386\ 406\ 102 \times 10^{22}$ s^{-1}
Toryx quantization constant	Λ	137
Rydberg constant	R_x	$1.097\ 373\ 157 \times 10^{07}$ m^{-1}
Planck length	l_p	$1.616252(81) \times 10^{-35}$ m

SUMMARY

Introduction

1. *Toryx* is a particular case of a multi-level, multi-dimensional spiral spacetime called *helicola* that permeates the Universe in the forms of spiral paths of celestial bodies winding around each other.
2. At each level of the helicola, there is a moving *leading string* and a *trailing string* winding around and moving synchronously with the leading string. The strings of adjacent levels overlap, so a trailing string of a previous level becomes a leading string of the next level.
3. The concept of spiral spacetime was introduced at the dawn of science almost 2600 years ago in the ancient Greece as a *vortex*. It then gradually gained its strength by absorbing the discoveries made in various branches of mathematics and sciences.
4. The toryces exist in both micro- and macro-worlds. The *micro-toryces* are the smallest entities of our Universe which *generic properties* can be found in many entities of the Universe. They form elementary matter particles. The *macro-toryces* are associated with celestial bodies and responsible for interactions between them.

Part 1 - Abstract Mathematics of a Toryx (Chapters 1 through 5)

1. Spacetime properties of a toryx are based on *three toryx spacetime postulates*. The first two postulates limit the degrees of freedom of its circular *leading string*, toroidal *trailing string* and its *eye radius*, while the third postulate requires the *spiral velocity of trailing string* to be equal to the velocity of light (Chapter 1).

2. Based on the three toryx spacetime postulates, it is necessary to modify three commonly-accepted aspects of elementary mathematics and physics(Chapter 2):

   a. Conventional zero (0) is replaced with *infinility* (± 0) that is an inverse of *infinity* ($\pm\infty$).
   b. Trigonometric cosine function between 180 and 360^0 is replaced with its inverse value.
   c. While spiral velocity of trailing string is constant and equal to the velocity of light, its either translational or rotational component *may exceed velocity of light*, causing the other component to be expressed with an imaginary number.

3. Toryces are topologically-polarized by their *vorticity* and *reality* (Chapters 2 and 3):

   - The *vorticity-polarized toryces* have opposite charges
   - The *reality-polarized toryces* are able to absorb and release spacetime (energy)
   - Symmetrical relationships between topologically-polarized toryces are conveniently described by circular diagrams containing both *infinity* and *infinility domains*
   - *Superluminal and subluminal components of spiral velocity* of toryx trailing string are respectively associated with *absorption and release of spacetime* by the toryx.

4. Trends of spacetime properties of toryces can be expressed as a function of steepness angle of toryx trailing string extending from 0^0 to 360^0 degrees that corresponds to the radius of toryx leading string extending from negative to positive infinity (Chapters 4 and 5).

Part 2 - Applied Mathematics of a Toryx (Chapters 6 through 10)

1. Toryces exist in two main kinds of quantum spacetime (energy) states (Chapter 6):

    *   *Excitation quantum states* at which only the radius of the toryx leading string increases, while its eye radius remains constant
    *   *Oscillation quantum states* at which both the radius of the toryx leading string and its eye radius change in the same proportion.

    The toryces in which their frequencies of trailing strings relate to one another by simple harmonic ratios are called *harmonic toryces.*

2. *Spacetime properties of toryces* are expressed in *objective spacetime units* comprehendible by all intelligent inhabitants in the Universe. Subjective *physical properties of toryces* invented by human beings are directly related to their respective *spacetime properties* (Chapters 7 and 8).

3. Toryces are formed from *quantum vacuum* in a form of harmonic toryces based on the *toryx uncertainty principle* (Chapter 7).

4. Four kinds of elementary matter particles, called *trons*, are formed by the unification of polarized toryces (Chapters 7 and 8):

    *   *Electrons* that model known electrons
    *   *Positrons* that model known positrons
    *   *Ethertrons* that model light elementary particles
    *   *Singulatrons* that model heavy elementary particles.

5. *Negative leptons* are merely the electrons in oscillation quantum states (Chapters 6 - 8).

6. All four elementary matter particles may exist at various matter levels Ls, besides the *ordinary matter level L2*. Stable trons follow the *tron polarization conservation law* assuring a balanced absorption and release of spacetime by their constituent toryces (Chapters 7 and 8).

7. Compositions of nucleons, atom and isotopes are (Chapter 9):

    *   *Nucleon core* is made up of exciting ethertrons and singulatrons residing at the vertices and a center of an *octahedral nucleon confinement* formed at the intersections of harmonic toryces.
    *   *Proton* contains a nucleon core and a harmonic positron located at the center of the nucleon core.
    *   *Stable neutron* is composed of a proton and a harmonic electron located at the center of the proton.
    *   *Unstable neutron* contains a proton and a resonant oscillated harmonic electron located at the center of the proton.
    *   *Nucleons of stable atoms* and *isotopes* are made up of protons and stable neutrons.

- *Nucleons of unstable atoms* and *isotopes* contain protons and at least one unstable neutron.
- *Hydrogen atom* is composed of a nucleon core, an excited positron and an excited electron.

8. *Macro-toryces* are the spacetime fields associated with celestial bodies and responsible for the interactions between them. Equations describing spacetime properties of micro- and macro-toryces are the same, except for the equations describing their eye radii (Chapters 1 and 10).

9. The macro-toryces obey two laws (Chapter 10):

- The *macro-toryx law of planetary motion* that for large distances between celestial bodies reduces to the Kepler's third law of planetary motion
- The *spacetime law of gravitation* that for large distances between celestial bodies reduces to the Newton's universal law of gravitation.

Confirmation of Measured Data

The values of calculated gravitational masses, charges and magnetic moments of several known particles, atoms and isotopes that are close to their measured values:

- Electron, muon and tau (see Table 8.22 and Fig. 8.22)
- Hydrogen atom (see Table 9.2.1)
- Proton (see Table 9.3.1)
- Unstable neutron (see Table 9.3.3)
- Deuterium (see Table 9.4.1)
- Tritium (see Table 9.4.2)
- Helium-3 (see Table 9.4.3)
- Helium-4 (see Table 9.4.4).

Predictions

The theory yields predicted masses, charges and magnetic moments of several unknown particles:

- Mass and magnetic moment of a still undiscovered stable neutron (see Table 9.3.2)

- Mass and magnetic moments of still undiscovered leptons (see Table 8.22 and Fig. 8.22):

$$3e \text{ lepton}............1.5199893 \text{ MeV}$$
$$Nu\ (\nu)\text{ lepton}.........18.2494 \text{ GeV}$$
$$Rho\ (\rho)..............131.8450 \text{ GeV}$$
$$Chi\ (\chi)\text{ lepton}......739.8507 \text{ GeV}.$$

Notably, the calculated mass of ***rho-tron*** is 5.5% greater than the mass of the Higgs boson of 125 GeV measured at CERN in 2017.

The Spacetime Origin of the Universe

Physical properties of toryces are directly related to their spacetime properties.

What current theories of physics describe as matter, field, charge, mass, electromagnetism and gravity are merely metamorphoses
of spiral spacetimes.

All spacetime parameters of micro-toryces and
elementary matter particles can be calculated based on
the following spacetime constants:

Velocity of light $c = 2.997\,924\,58 \times 10^{08}\,\text{m/s}$

Micro-toryx eye radius $r_0 = 1.408\,970\,164 \times 10^{-15}\,\text{m}$

Toryx quantization constant $\Lambda = 137$.

<u>The rest is commentary, as described in this book.</u>

Appendix A

DERIVATION OF TORYX EQUATIONS

CONTENTS

A1 Toryx Parameters as Functions of b_1

A1.1 Toryx Radii

Eq. (1.6-1a)
Relative radius of toryx leading string – From Eq. (1.4-2):

$$b_1 = \frac{r_1}{r_0} \tag{A1.1-1}$$

Eq. (1.6-1b)
Relative radius of toryx trailing string – From Eqs. (1.4-3) and (1.5-2):

$$b_2 = \frac{r_2}{r_0} = b_1 - 1 \tag{A1.1-2}$$

Eq. (1.6-13)
Relative radius of toryx spherical boundary – From Eqs. (A1.1-1) and Fig. 1.1.2:

$$b = \frac{r}{r_0} = \frac{2r_1 - r_0}{r_0} = 2b_1 - 1$$

$$b = 2b_1 - 1 \tag{A1.1-3}$$

Eq. (6.5-2)
Relative radius of toryx fine-structure spherical boundary – From Eqs. (6.5-1), (A1.1-1) and Fig. (6.5.1):

$$b_3 = \frac{r_3}{r_0} = \frac{\sqrt{r_1^2 - r_2^2}}{r_0} = \frac{\sqrt{r_1^2 - (r_1 - r_0)^2}}{r_0} = \frac{\sqrt{2r_1 - r_0}}{r_0} = \sqrt{2b_1 - 1}$$

$$b_3 = \sqrt{2b_1 - 1} \qquad\qquad\qquad\qquad (A1.1\text{-}4)$$

A1.2 Toryx Lengths, Wavelengths & Number of Windings

Eq. (1.6-3a)
Relative length of one winding of toryx leading string – From Eqs. (1.4-4) and (A1.1-1):

$$l_1 = \frac{L_1}{2\pi r_0} = \frac{2\pi r_1}{2\pi r_0} = b_1$$

$$l_1 = \frac{L_1}{2\pi r_0} = b_1 \qquad\qquad\qquad\qquad (A1.2\text{-}1)$$

Eq. (1.6-3b)
Relative length of one winding of toryx trailing string – From Eqs. (1.3-1) and (1.4-5):

$$l_2 = \frac{L_2}{2\pi r_0} = \frac{2\pi r_1}{2\pi r_0} = b_1$$

$$l_2 = \frac{L_2}{2\pi r_0} = b_1 \qquad\qquad\qquad\qquad (A1.2\text{-}2)$$

Eq. (1.6-2b)
Relative wavelength of toryx trailing string – From Eqs. (1.4-17), (A1.1-1), (A1.1-2) and Fig. 1.3:

$$\eta_2 = \frac{\lambda_2}{2\pi r_0} = \frac{\sqrt{(2\pi r_1)^2 - (2\pi r_2)^2}}{2\pi r_0} = \sqrt{b_1^2 - b_2^2} = \sqrt{b_1^2 - (b_1 - 1)^2} = \sqrt{2b_1 - 1}$$

$$\eta_2 = \frac{\lambda_2}{2\pi r_0} = \sqrt{2b_1 - 1} \qquad\qquad\qquad\qquad (A1.2\text{-}3)$$

Eq. (1.6-5b)
The number of widings of toryx trailing string – From Eqs. (A1.2-1) and (A1.2-3):

$$w_2 = \frac{L_1}{\lambda_2} = \frac{2\pi b_1 r_0}{2\pi r_0 \sqrt{2b_1 - 1}} = \frac{b_1}{\sqrt{2b_1 - 1}}$$

$$w_2 = \frac{b_1}{\sqrt{2b_1 - 1}} \qquad\qquad\qquad\qquad (A1.2\text{-}4)$$

A1.3 Steepness Angle of Trailing String

Eq. (1.6-4b)
- **Cosine of steepness angle of toryx trailing string** – From Eqs. (1.6-11), (1.6-12) and Fig. 1.5:

$$\cos s\varphi_2 = \frac{b_1 - 1}{b_1} \tag{A1.3-1}$$

- **Sine of steepness angle of toryx trailing string** – From Eq. (A1.3-1):

$$\sin s\varphi_2 = \sqrt{1 - \cos s^2\varphi_2} = \sqrt{1 - \left(\frac{b_1 - 1}{b_1}\right)^2} = \frac{\sqrt{2b_1 - 1}}{b_1}$$

$$\sin s\varphi_2 = \frac{\sqrt{2b_1 - 1}}{b_1} \tag{A1.3-2}$$

- **Tangent of steepness angle of toryx trailing string** – From Eqs. (A1.3-1) and (A1.3-2):

$$\tan s\varphi_2 = \frac{\sin s\varphi_2}{\cos s\varphi_2} = \frac{\sqrt{2b_1 - 1}}{b_1} \frac{b_1}{b_1 - 1} = \frac{\sqrt{2b_1 - 1}}{b_1 - 1}$$

$$\tan s\varphi_2 = \frac{\sqrt{2b_1 - 1}}{b_1 - 1} \tag{A1.3-3}$$

A1.4 Middle Velocities of Toryx Strings (Fig. 1.1.3, point *a*)

Eq. (1.6-7b)
Relative rotational velocity of toryx trailing string – From Eqs. (1.4-13), (1.5-3), (A1.3-1) and Fig. 1.5:

$$\beta_{2r} = \frac{V_{2r}}{c} = \beta_2 \cos s\varphi_2 = \frac{b_1 - 1}{b_1}$$

$$\beta_{2r} = \frac{V_{2r}}{c} = \frac{b_1 - 1}{b_1} \tag{A1.4-1}$$

Eq. (1.6-6b)
Relative translational velocity of toryx trailing string – From Eqs. (1.4-12), (1.5-3), (A1.4-1), and Fig. 1.5:

$$\beta_{2t} = \frac{V_{2t}}{c} = \sqrt{1 - \beta_{2r}^2} = \sqrt{1 - \left(\frac{b_1 - 1}{b_1}\right)^2} = \frac{\sqrt{2b_1 - 1}}{b_1}$$

$$\beta_{2t} = \frac{V_{2t}}{c} = \frac{\sqrt{2b_1 - 1}}{b_1} \qquad (A1.4\text{-}2)$$

Eqs. (1.6-7a) and (1.6-8a)
Relative spiral and rotational velocities of toryx leading string – From Eqs. (1.1-4), (1.4-8), (1.4-10) and (A1.4-2):

$$\beta_1 = \beta_{1r} = \beta_{2t} = \frac{\sqrt{2b_1 - 1}}{b_1}$$

$$\beta_1 = \frac{V_1}{c} = \beta_{1r} = \frac{V_{1r}}{c} = \frac{\sqrt{2b_1 - 1}}{b_1} \qquad (A1.4\text{-}3)$$

A1.5 Frequencies of Toryx Strings

Eq. (1.6-9a)
Relative frequency of toryx leading string – From Eqs. (1.4-14), (1.4-18), (A1.2-1) and (A1.4-3):

$$\delta_1 = \frac{f_1}{f_0} = \frac{V_1}{L_1 f_0} = \frac{V_1}{c}\frac{c}{2\pi r_0}\frac{1}{b_1 f_0} = \frac{\beta_1}{b_1} = \frac{\sqrt{2b_1 - 1}}{b_1^2}$$

$$\delta_1 = \frac{f_1}{f_0} = \frac{\sqrt{2b_1 - 1}}{b_1^2} \qquad (A1.5\text{-}1)$$

Eq. (1.6-9b)
Relative frequency of toryx trailing string – From Eqs. (1.3-1), (1.5-3), (1.4-11), (1.4-18) and (A1.2-2):

$$\delta_2 = \frac{f_2}{f_0} = \frac{V_2}{L_2 f_0} = \frac{c}{2\pi r_0}\frac{\beta_2}{b_1 f_0} = \frac{1}{b_1}$$

$$\delta_2 = \frac{f_2}{f_0} = \frac{1}{b_1} \qquad (A1.5\text{-}2)$$

A1.6 Peripheral Velocities of Trailing String (Fig. 3.3.1)

Eq. (3.3-1)
Relative inner translational velocity of trailing string - From Eq. (A1.4-2):

$$\beta_{2t}^{in} = \beta_{2t}\frac{1}{b_1} = \frac{\sqrt{2b_1-1}}{b_1^2}$$

$$\beta_{2t}^{in} = \frac{\sqrt{2b_1-1}}{b_1^2} \tag{A1.6-1}$$

Eq. (3.3-2)
Relative outer translational velocity of trailing string - From Eq. (A1.4-2):

$$\beta_{2t}^{out} = \beta_{2t}\frac{2b_1-1}{b_1} = \frac{\sqrt{2b_1-1}(2b_1-1)}{b_1^2} = \frac{(2b_1-1)^{1.5}}{b_1^2}$$

$$\beta_{2t}^{out} = \frac{(2b_1-1)^{1.5}}{b_1^2} \tag{A1.6-2}$$

Eq. (3.3-3)
Relative inner rotational velocity of trailing string - From Eqs. (1.5-3) and (A1.6-1):

$$\beta_{2r}^{in} = \sqrt{1-\left(\beta_{2t}^{in}\right)^2} = \sqrt{1-\frac{2b_1-1}{b_1^4}} = \frac{\sqrt{b_1^4-2b_1+1}}{b_1^2}$$

$$\beta_{2r}^{in} = \frac{\sqrt{b_1^4-2b_1+1}}{b_1^2} \tag{A1.6-3}$$

Eq. (3.3-4)
<u>**Relative outer rotational velocity of trailing string**</u> - From Eqs. (1.5-3) and (A1.6-2):

$$\beta_{2r}^{out} = \sqrt{1-\left(\beta_{2t}^{out}\right)^2} = \sqrt{1-\frac{(2b_1-1)^3}{b_1^4}} = \frac{\sqrt{b_1^4-(2b_1-1)^3}}{b_1^2}$$

$$\beta_{2r}^{out} = \frac{\sqrt{b_1^4-(2b_1-1)^3}}{b_1^2} \tag{A1.6-4}$$

A2 Toryx Parameters as Function of φ_2 (Fig. 1.5)

A2.1 Toryx Radii

Eq. (1.6-14a)
Relative radius of toryx leading string – From Eq. (A1.3-1):

$$\cos s\varphi_2 = \frac{b_1 - 1}{b_1} = 1 - \frac{1}{b_1}; \quad \frac{1}{b_1} = 1 - \cos s\varphi_2$$

$$b_1 = \frac{1}{1 - \cos s\varphi_2} \tag{A2.1-1}$$

Eq. (1.6-14b)
Relative radius of toryx trailing string – From Eqs. (A1.1-2) and (A2.1-1):

$$b_2 = b_1 - 1 = \frac{1}{1 - \cos s\varphi_2} - 1 = \frac{1 - 1 + \cos s\varphi_2}{1 - \cos s\varphi_2} = \frac{\cos s\varphi_2}{1 - \cos s\varphi_2}$$

$$b_2 = \frac{\cos s\varphi_2}{1 - \cos s\varphi_2} \tag{A2.1-2}$$

Eq. (1.6-23)
Relative toryx spherical boundary radius – From Eqs. (A1.1-3) and (A2.1-1):

$$b = 2b_1 - 1 = \frac{2}{1 - \cos s\varphi_2} - 1 = \frac{1 + \cos s\varphi_2}{1 - \cos s\varphi_2}$$

$$b = \frac{1 + \cos s\varphi_2}{1 - \cos s\varphi_2} \tag{A2.1-3}$$

A2.2 Toryx Lengths, Wavelengths & Number of Windings

Eq. (1.6-16a) and (1.6-16b)
Relative length of one winding of toryx leading and trailing strings – From Eqs. (A1.2-1), (A1.2-2) and (A2.1-1):

$$l_1 = l_2 = \frac{1}{1 - \cos s\varphi_2} \tag{A2.2-1}$$

Eq. (1.6-15b)
Relative wavelength of toryx trailing string – From Eqs. (A1.2-3) and (A2.1-1):

$$\eta_2 = \sqrt{2b_1 - 1} = \sqrt{\frac{2}{1 - \cos s\varphi_2} - 1} = 2\sqrt{\frac{1 + \cos s\varphi_2}{1 - \cos s\varphi_2}}$$

$$\eta_2 = \sqrt{\frac{1 + \cos s\varphi_2}{1 - \cos s\varphi_2}} \tag{A2.2-2}$$

Eq. (1.6-17b)
The number of windings of toryx trailing string – From Eqs. (A1.2-4) and (A2.1-1):

$$w_2 = \frac{b_1}{\sqrt{2b_1 - 1}} = \frac{1}{1 - \cos s\varphi_2} \frac{1}{\sqrt{\dfrac{2}{1 - \cos s\varphi_2} - 1}} = \frac{1}{1 - \cos s\varphi_2} \frac{1}{\sqrt{\dfrac{1 + \cos s\varphi_2}{1 - \cos s\varphi_2}}} =$$

$$\frac{1}{\sqrt{\dfrac{(1 - \cos s\varphi_2)^2 (1 + \cos s\varphi_2)}{1 - \cos s\varphi_2}}} = \frac{1}{\sqrt{1 - \cos s^2\varphi_2}} = \frac{1}{\sqrt{\sin s^2\varphi_2}} = \frac{1}{\sin s\varphi_2}$$

$$w_2 = \frac{1}{\sin s\varphi_2} \tag{A2.2-3}$$

A2.3 Middle Velocities of Toryx Strings (Fig. 1.5)

Eq. (1.6-18b)
Relative translational velocity of toryx trailing string - From Eqs. (1.5-3) and (1.6-8b):

$$\beta_{2t} = \beta_2 \sin s\varphi_2 = \sin s\varphi_2$$
$$\beta_{2t} = \sin s\varphi_2 \tag{A2.3-1}$$

Eqs. (1.6-19a) and (1.6-20a)
Relative spiral and rotational velocities of toryx leading string – From Eqs. (A1.4-3) and (A2.3-1):

$$\beta_1 = \beta_{1r} = \beta_{2t} = \sin s\varphi_2$$
$$\beta_1 = \beta_{1r} = \sin s\varphi_2 \tag{A2.3-2}$$

Eq. (1.6-19b)
Relative rotational velocity of toryx trailing string – From Eq. (1.5-3):

$$\beta_{2r} = \beta_2 \cos s\varphi_2 = \cos s\varphi_2$$
$$\beta_{2r} = \cos s\varphi_2 \tag{A2.3-3}$$

A2.4 Frequencies of Toryx Strings

Eq. (1.6-21a)
Relative frequency of toryx leading string – From Eqs. (A1.4-2), (A1.5-1), (A2.1-1) and (A2.3-1):

$$\delta_1 = \frac{\sqrt{2b_1 - 1}}{b_1^2} = \frac{\beta_{2t}}{b_1} = \sin s\varphi_2 (1 - \cos s\varphi_2)$$

$$\beta_{2t} = \sin s\varphi_2 (1 - \cos s\varphi_2) \qquad\qquad \text{(A2.4-1)}$$

Eq. (1.6-21b)
Relative frequency of toryx trailing string – From Eqs. (A1.5-2) and (A2.1-1):

$$\delta_2 = \frac{1}{b_1} = 1 - \cos s\varphi_2$$

$$\delta_2 = 1 - \cos s\varphi_2 \qquad\qquad \text{(A2.4-2)}$$

A3 Relationship between Parameters of Toryces of Main Groups

B3.1 Radii of Reality-Polarized Toryces (Fig. 2.3.1)

Eq. (3.2-5a)
Relative radius of leading string of reality-polarized negative toryces $\breve{E}^- \leftrightarrow E^-$ - From Eqs. (2.4-1) and (3.2-1a):

$$\breve{V}_E^- = \frac{1}{V_E^-}; \quad -\frac{\breve{b}_{1E}^- - 1}{\breve{b}_{1E}^-} = -\frac{b_{1E}^-}{b_{1E}^- - 1}; \quad -1 + \frac{1}{\breve{b}_{1E}^-} = -\frac{b_{1E}^-}{b_{1E}^- - 1}; \quad \frac{1}{\breve{b}_{1E}^-} = -\frac{b_{1E}^-}{b_{1E}^- - 1} + 1;$$

$$\frac{1}{\breve{b}_{1E}^-} = \frac{-b_{1E}^- + b_{1E}^- - 1}{b_{1E}^- - 1} = \frac{-1}{b_{1E}^- - 1} = \frac{1}{1 - b_{1E}^-}$$

$$\breve{b}_{1E}^- = 1 - b_{1E}^- \qquad\qquad \text{(A3.1-1a)}$$

- **Relative radius of leading string of reality-polarized negative toryces** $\breve{A}^- \leftrightarrow A^-$ - Similarly to Eq. (A3.1-1a):

$$\breve{b}_{1A}^- = 1 - b_{1A}^- \qquad\qquad \text{(A3.1-1b)}$$

Eq. (3.2-6a)
Relative radius of leading string of reality-polarized positive toryces $\breve{E}^+ \leftrightarrow E^+$ - From Eqs.
(2.4-1) and (3.2-2a):

$$\breve{V}_E^+ = \frac{1}{V_E^+}; \quad \frac{\breve{b}_{1E}^+ - 1}{\breve{b}_{1E}^+} = \frac{b_{1E}^+}{b_{1E}^+ - 1}; \quad 1 - \frac{1}{\breve{b}_{1E}^+} = \frac{b_{1E}^+}{b_{1E}^+ - 1}; \quad 1 - \frac{b_{1E}^+}{b_{1E}^+ - 1} = \frac{1}{\breve{b}_{1E}^+};$$

$$\frac{b_{1E}^+ - 1 - b_{1E}^+}{b_{1E}^+ - 1} = \frac{1}{\breve{b}_{1E}^+}; \quad \frac{-1}{b_{1E}^+ - 1} = \frac{1}{\breve{b}_{1E}^+}; \quad \breve{b}_{1E}^+ = 1 - b_{1E}^+$$

$$\breve{b}_{1E}^+ = 1 - b_{1E}^+ \tag{A3.1-2a}$$

- **Relative radius of leading string of reality-polarized positive toryces** $\breve{A}^+ \leftrightarrow A^+$ - Similarly to Eq. (A3.1-2a):

$$\breve{b}_{1A}^+ = 1 - b_{1A}^+ \tag{A3.1-2b}$$

Eq. (3.2-5b)
Relative radius of spherical boundary of reality-polarized negative toryces $\breve{E}^- \leftrightarrow E^-$ - From Eqs. (1.6-13) and (A3.1-1a):

$$\breve{b}_E^- = 2\breve{b}_{1E}^- - 1 = 2(1 - b_{1E}^-) - 1 = 2 - 2b_{1E}^- - 1 = 1 - 2b_{1E}^- =$$

$$1 - 2\frac{b_E^- + 1}{2} = 1 - b_E^- - 1 = -b_E^-$$

$$\breve{b}_E^- = -b_E^- \tag{A3.1-3a}$$

- **Relative radius of spherical boundary of reality-polarized negative toryces** $\breve{A}^- \leftrightarrow A^-$ -
  Similarly to Eq. (A3.1-3a):

$$\breve{b}_A^- = -b_A^- \tag{A3.1-3b}$$

Eq. (3.2-6b)
Relative radius of spherical boundary of reality-polarized positive toryces $\breve{E}^+ \leftrightarrow E^+$ - From Eqs. (1.6-13) and (A3.1-2a):

$$\breve{b}_E^+ = 2\breve{b}_{1E}^+ - 1 = 2(1 - b_{1E}^+) - 1 = 2 - 2b_{1E}^+ - 1 = 1 - 2b_{1E}^+ =$$

$$1 - 2\frac{b_E^+ + 1}{2} = 1 - b_E^+ - 1 = -b_E^+$$

$$\breve{b}_E^+ = -b_E^+ \tag{A3.1-4a}$$

- **Relative radius of spherical boundary of reality-polarized positive toryces** $\breve{A}^+ \leftrightarrow A^+$ - Similarly to Eq. (A3.1-4a):

$$\breve{b}_A^+ = -b_A^+ \tag{A3.1-4b}$$

A3.2 Radii of Vorticity-Polarized Toryces (Fig. 2.3.1)

Eq. (3.2-7a)
Relative radius of leading string of vorticity-polarized real toryces $A^+ \leftrightarrow A^-$ - From Eqs. (1.6-13), (2.4-1) and (3.2-3a):

$$V_A^+ = -V_A^-; \quad -\frac{b_{1A}^+ - 1}{b_{1A}^+} = \frac{b_{1A}^- - 1}{b_{1A}^-}; \quad -1 + \frac{1}{b_{1A}^+} = 1 - \frac{1}{b_{1A}^-}; \quad \frac{1}{b_{1A}^+} = 2 - \frac{1}{b_{1A}^-};$$

$$\frac{1}{b_{1A}^+} = \frac{2b_{1A}^- - 1}{b_{1A}^-}; \quad b_{1A}^+ = \frac{b_{1A}^-}{2b_{1A}^- - 1}$$

$$b_{1A}^+ = \frac{b_{1A}^-}{2b_{1A}^- - 1} \tag{A3.2-1a}$$

- **Relative radius of leading string of vorticity-polarized real toryces** $E^+ \leftrightarrow E^-$ - Similarly to Eq. (A3.2-1a):

$$b_{1E}^+ = \frac{b_{1E}^-}{2b_{1E}^- - 1} \tag{A3.2-1b}$$

Eq. (3.2-8a)
Relative radius of leading string of vorticity-polarized imaginary toryces $\breve{A}^+ \leftrightarrow \breve{A}^-$ - From Eqs. (2.4-1), (3.2-4a):

$$\breve{V}_A^+ = -\breve{V}_A^-; \quad -\frac{\breve{b}_{1A}^+ - 1}{\breve{b}_{1A}^+} = \frac{\breve{b}_{1A}^- - 1}{\breve{b}_{1A}^-}; \quad -1 + \frac{1}{\breve{b}_{1A}^+} = 1 - \frac{1}{\breve{b}_{1A}^-}; \quad \frac{1}{\breve{b}_{1A}^+} = 2 - \frac{1}{\breve{b}_{1A}^-};$$

$$\frac{1}{\breve{b}_{1A}^+} = \frac{2\breve{b}_{1A}^- - 1}{\breve{b}_{1A}^-}; \quad \breve{b}_{1A}^+ = \frac{\breve{b}_{1A}^-}{2\breve{b}_{1A}^- - 1}$$

$$\breve{b}_{1A}^+ = \frac{\breve{b}_{1A}^-}{2\breve{b}_{1A}^- - 1} \tag{A3.2-1a}$$

- **Relative radius of leading string of vorticity-polarized imaginary toryces** $\breve{E}^+ \leftrightarrow \breve{E}^-$ - Similarly to Eq. (A3.2-1a):

$$\breve{b}_{1E}^+ = \frac{\breve{b}_{1E}^-}{2\breve{b}_{1E}^- - 1} \tag{A3.2-1b}$$

Eq. (3.2-7b)
Relative radius of spherical boundary of vorticity-polarized real toryces $A^+ \leftrightarrow A^-$ - From
Eqs. (1.6-13) and (A3.2-1a):

$$b_A^+ = 2b_{1A}^+ - 1 = \frac{2b_{1A}^-}{2b_{1A}^- - 1} - 1 = \frac{2b_{1A}^- - 2b_{1A}^- + 1}{2b_{1A}^- - 1} = \frac{1}{2b_{1A}^- - 1} = \frac{1}{b_A^-}$$

$$b_A^+ = \frac{1}{b_A^-} \qquad\qquad (A3.2\text{-}2a)$$

- **Relative radius of spherical boundary of vorticity-polarized real toryces** $E^+ \leftrightarrow E^-$ -
  Similarly to Eq. (A3.2-2a):

$$b_E^+ = \frac{1}{b_E^-} \qquad\qquad (A3.2\text{-}2b)$$

Eq. (3.2-8b)
Relative radius of spherical boundary of vorticity-polarized imaginary toryces $\breve{A}^+ \leftrightarrow \breve{A}^-$ -
From Eqs. (1.6-13) and (A3.2-1a):

$$\breve{b}_A^+ = 2\breve{b}_{1A}^+ - 1 = \frac{2\breve{b}_{1A}^-}{2\breve{b}_{1A}^- - 1} - 1 = \frac{2\breve{b}_{1A}^- - 2\breve{b}_{1A}^- + 1}{2\breve{b}_{1A}^- - 1} = \frac{1}{2\breve{b}_{1A}^- - 1} = \frac{1}{\breve{b}_A^-}$$

$$\breve{b}_A^+ = \frac{1}{\breve{b}_A^-} \qquad\qquad (A3.2\text{-}3a)$$

- **Relative radius of spherical boundary of vorticity-polarized real toryces** $\breve{E}^+ \leftrightarrow \breve{E}^-$ -
  Similarly to Eq. (A3.2-3a):

$$\breve{b}_E^+ = \frac{1}{\breve{b}_E^-} \qquad\qquad (A3.2\text{-}3b)$$

A4 Relationship between Parameters of Toryces of Subgroups

Eq. (3.2-14a)
Relative radius of leading string of real negative toryces $A^- \leftrightarrow E^-$ - From Eqs. (2.4-1) and
(3.2-10):

$$V_A^- + V_E^- = -1; \quad \frac{b_{1A}^- - 1}{b_{1A}^-} + \frac{b_{1E}^- - 1}{b_{1E}^-} = 1; \quad 1 - \frac{1}{b_{1A}^-} + 1 - \frac{1}{b_{1E}^-} = 1; \quad \frac{1}{b_{1A}^-} = 1 - \frac{1}{b_{1E}^-};$$

$$\frac{1}{b_{1A}^-} = \frac{b_{1E}^- - 1}{b_{1E}^-}; \quad b_{1A}^- = \frac{b_{1E}^-}{b_{1E}^- - 1}$$

$$b_{1A}^{-} = \frac{b_{1E}^{-}}{b_{1E}^{-} - 1}$$

(A4.1-1)

Eq. (3.2-15a)

Relative radius of leading string of real positive toryces $A^{+} \leftrightarrow E^{+}$ - From Eqs. (2.4-1) and (3.2-11):

$$V_A^{+} + V_E^{+} = 1; \quad -\frac{b_{1A}^{+}-1}{b_{1A}^{+}} - \frac{b_{1E}^{+}-1}{b_{1E}^{+}} = 1; \quad -1 + \frac{1}{b_{1A}^{+}} - 1 + \frac{1}{b_{1E}^{+}} = 1; \quad \frac{1}{b_{1A}^{+}} = 3 - \frac{1}{b_{1E}^{+}};$$

$$\frac{1}{b_{1A}^{+}} = \frac{3b_{1E}^{+}-1}{b_{1E}^{+}}; \quad b_{1A}^{+} = \frac{b_{1E}^{+}}{3b_{1E}^{+}-1}$$

$$b_{1A}^{+} = \frac{b_{1E}^{+}}{3b_{1E}^{+}-1}$$

(A4.1-2)

Eq. (3.2-14b)

Relative radius of spherical boundary of real negative toryces $A^{-} \leftrightarrow E^{-}$ - From Eqs. (1.6-13) and (A4.1-1):

$$b_A^{-} = 2b_{1A}^{-} - 1 = \frac{2b_{1E}^{-}}{b_{1E}^{-}-1} - 1 = \frac{2(b_E^{-}+1)}{2} \frac{1}{\frac{b_E^{-}+1}{2}-1} - 1 = \frac{2(b_E^{-}+1)}{b_E^{-}+1-2} - 1 = \frac{2(b_E^{-}+1)}{b_E^{-}-1} - 1;$$

$$\frac{2b_E^{-}+2}{b_E^{-}-1} - 1 = \frac{2b_E^{-}+2-b_E^{-}+1}{b_E^{-}-1} = \frac{b_E^{-}+3}{b_E^{-}-1}$$

$$b_A^{-} = \frac{b_E^{-}+3}{b_E^{-}-1}$$

(A4.1-3)

Eq. (3.2-15b)

Relative radius of spherical boundary of real positive toryces $A^{+} \leftrightarrow E^{+}$ - From Eqs. (1.6-13) and (A4.1-2):

$$b_A^{+} = 2b_{1A}^{+} - 1 = \frac{2b_{1E}^{+}}{3b_{1E}^{+}-1} - 1 = \frac{2(b_E^{+}+1)}{2} \frac{1}{\frac{3(b_E^{+}+1)}{2}-1} - 1 = \frac{2(b_E^{+}+1)}{3b_E^{+}+3-2} - 1 = \frac{2b_E^{+}+2}{3b_E^{+}+1} - 1;$$

$$\frac{2b_E^{+}+2}{3b_E^{+}+1} - 1 = \frac{2b_E^{+}+2-3b_E^{+}-1}{3b_E^{+}+1} = \frac{-b_E^{+}+1}{3b_E^{+}+1}$$

$$b_A^{+} = \frac{1-b_E^{+}}{1+3b_E^{+}}$$

(A4.1-4)

A5 Excitation Quantum States of Toryces

Eq. (6.2-1) - From Eq. (6.1-1):

$$b_{1E}^- = z \qquad\qquad\text{(A5-1)}$$

Eq. (6.2-2) - From Eqs. (A5-1) and (1.6-13):

$$b_E^- = 2b_E^- - 1 = 2z - 1$$
$$b_E^- = 2z - 1 \qquad\qquad\text{(A5-2)}$$

Eq. (6.2-3) - From Eqs. (A3.1-1a) and (A5-1):

$$\breve{b}_{1E}^- = 1 - b_{1E}^- = 1 - z$$
$$b_{1E}^- = 1 - z \qquad\qquad\text{(A5-3)}$$

Eq. (6.2-4) - From Eqs. (A3.1-3a) and (A5-2):

$$\breve{b}_E^- = -b_E^- = 1 - 2z$$
$$\breve{b}_E^- = 1 - 2z \qquad\qquad\text{(A5-4)}$$

Eq. (6.2-5) - From Eqs. (A3.2-1b) and (A5-1):

$$b_{1E}^+ = \frac{b_{1E}^-}{2b_{1E}^- - 1} = \frac{z}{2z - 1}$$

$$b_{1E}^+ = \frac{z}{2z - 1} \qquad\qquad\text{(A5-5)}$$

Eq. (6.2-6) - From Eqs. (A3.2-2b) and (A5-2):

$$b_E^+ = \frac{1}{b_E^-} = \frac{1}{2z - 1}$$

$$b_E^- = \frac{1}{2z - 1} \qquad\qquad\text{(A5-6)}$$

Eq. (6.2-7) - From Eqs. (A3.1-2a) and (A5-5):

$$\breve{b}_{1E}^{+} = 1 - b_{1E}^{+} = 1 - \frac{z}{2z-1} = \frac{2z-1-z}{2z-1} = \frac{z-1}{2z-1} = \frac{1-z}{1-2z}$$

$$b_{1E}^{+} = \frac{1-z}{1-2z} \qquad (A5\text{-}7)$$

Eq. (6.2-8) - From Eqs. (A3.2-3b) and (A5-4):

$$\breve{b}_{E}^{+} = \frac{1}{\breve{b}_{E}^{-}} = \frac{1}{1-2z}$$

$$\breve{b}_{E}^{+} = \frac{1}{1-2z} \qquad (A5\text{-}8)$$

Eq. (6.2-9) - From Eqs. (A4.1-1) and (A5-1):

$$b_{1A}^{-} = \frac{b_{1E}^{-}}{b_{1E}^{-} - 1} = \frac{z}{z-1}$$

$$b_{1A}^{-} = \frac{z}{z-1} \qquad (A5\text{-}9)$$

Eq. (6.2-10) - From Eqs. (1.6-13) and (A5-9):

$$b_{A}^{-} = 2b_{1A}^{-} - 1 = \frac{2z}{z-1} - 1 = \frac{2z-z+1}{z-1} = \frac{z+1}{z-1}$$

$$b_{A}^{-} = \frac{z+1}{z-1} \qquad (A5\text{-}10)$$

Eq. (6.2-11) - From Eqs. (A3.2-1a) and (A5-9):

$$b_{1A}^{+} = \frac{b_{1A}^{-}}{2b_{1A}^{-} - 1} = \frac{z}{z-1} \frac{1}{\dfrac{2z}{z-1} - 1} = \frac{z}{2z-z+1} = \frac{z}{z+1}$$

$$b_{1A}^{+} = \frac{z}{z+1} \qquad (A5\text{-}11)$$

Eq. (6.2-12) - From Eqs. (A3.2-2a) and (A5-10):

$$b_A^+ = \frac{1}{b_A^-} = \frac{z-1}{z+1}$$

$$b_A^+ = \frac{z-1}{z+1} \tag{A5-12}$$

Eq. (6.2-13) - From Eqs. (A3.1-1b) and (A5-9):

$$\breve{b}_{1A}^- = 1 - b_{1A}^- = 1 - \frac{z}{z-1} = \frac{z-1-z}{z-1} = \frac{-1}{z-1} = \frac{1}{1-z}$$

$$\breve{b}_{1A}^- = \frac{1}{1-z} \tag{A5-13}$$

Eq. (6.2-14) - From Eqs. (1.6-13) and (A5-13):

$$\breve{b}_A^- = 2\breve{b}_{1A}^- - 1 = \frac{2}{1-z} - 1 = \frac{2-1+z}{1-z} = \frac{1+z}{1-z}$$

$$\breve{b}_A^- = \frac{1+z}{1-z} \tag{A5-14}$$

Eq. (6.2-15) - From Eqs. (A3.2-1a) and (A5-13):

$$\breve{b}_{1A}^+ = \frac{\breve{b}_{1A}^-}{2\breve{b}_{1A}^- - 1} = \frac{1}{1-z} \frac{1}{\dfrac{2}{1-z} - 1} = \frac{1}{2-1+z} = \frac{1}{1+z}$$

$$\breve{b}_{1A}^+ = \frac{1}{1+z} \tag{A5-15}$$

Eq. (6.2-16) - From Eqs. (A3.2-3a) and (A5-14):

$$\breve{b}_A^+ = \frac{1}{\breve{b}_A^-} = \frac{1-z}{1+z}$$

$$\breve{b}_A^+ = \frac{1-z}{1+z} \tag{A5-16}$$

A6 Toryx Reality Ratio

Eq. (2.4-1)
Vorticity of real toryces – By definition:

$$V = -\frac{b_1 - 1}{b_1} \tag{A6-1}$$

- **Vorticity of imaginary toryces** – From Eqs. (A6-1), (3.2-5a) and (3.2-6a):

$$\breve{V} = -\frac{\breve{b}_1 - 1}{\breve{b}_1} = -\frac{1 - b_1 - 1}{1 - b_1} = -\frac{b_1}{b_1 - 1}$$

$$\breve{V} = -\frac{b_1}{b_1 - 1} \tag{A6-2}$$

Eq. (7.4-3) – **Toryx reality ratio** – From Eqs. (A6-1) and (A6-2):

$$T = \frac{\breve{V}}{V} = \frac{N}{\breve{N}} = \left(-\frac{b_1}{b_1 - 1}\right)\left(-\frac{b_1}{b_1 - 1}\right) = \left(\frac{b_1}{b_1 - 1}\right)^2$$

$$T = \frac{\breve{V}}{V} = \frac{N}{\breve{N}} = \left(\frac{b_1}{b_1 - 1}\right)^2 \tag{A6-3}$$

A7 Instant Velocities of Trailing String (Fig. 3.3.3)

- Increment of relative distance Δb along axis X within trailing string:

$$\Delta b = \frac{2(b_1 - 1)}{m} \tag{A7-1}$$

- X-coordinate x_m of middle point m of increment z_m:

$$x_m = x_{m-1} + \Delta b \tag{A7-2}$$

- X-coordinate x_{m1} of entry point m_1 of increment z_m:

$$x_{m1} = x_m - \frac{\Delta b}{2} \tag{A7-3}$$

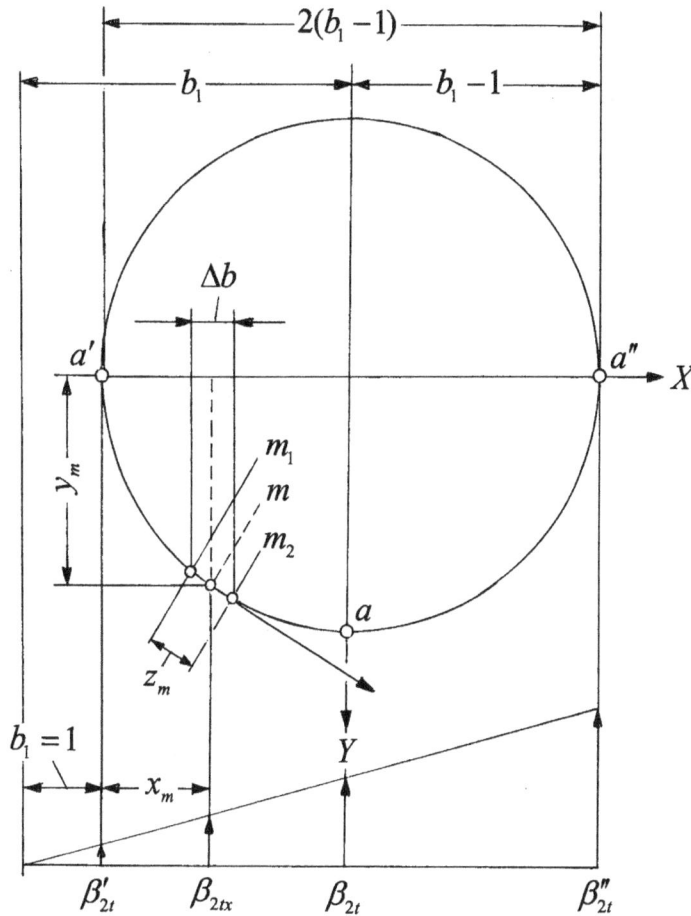

Figure Fig. 3.3.3. Derivation of instantaneous velocities of trailing string.

- Increment of relative distance Δb along axis X within trailing string:

$$\Delta b = \frac{2(b_1 - 1)}{m} \qquad \text{(A7-1)}$$

- X-coordinate x_m of middle point m of increment z_m:

$$x_m = x_{m-1} + \Delta b \qquad \text{(A7-2)}$$

- X-coordinate x_{m1} of entry point m_1 of increment z_m:

$$x_{m1} = x_m - \frac{\Delta b}{2} \qquad \text{(A7-3)}$$

- X-coordinate x_{m2} of exit point m_2 of increment z_m:

$$x_{m2} = x_m + \frac{\Delta b}{2} \tag{A7-4}$$

- Y-coordinate y_{m1} of entry point m_1 of increment z_m:

$$y_{m1} = \sqrt{x_{m1}[2(b_1 - 1) - x_{m1}]} \tag{A7-5}$$

- Y-coordinate y_{m2} of exit point m_2 of increment z_m:

$$y_{m2} = \sqrt{x_{m1}[2(b_1 - 1) - x_{m2}]} \tag{A7-6}$$

- Increment z_m:

$$z_m = \sqrt{(\Delta b)^2 + (y_{m2} - y_{m1})^2} \tag{A7-7}$$

- Relative translational velocity β_{2tx} of point m - From Eq. (1.6-6b):

$$\beta_{2tx} = \beta_{2t} \frac{x_m + 1}{b_1} = (x_m + 1) \frac{\sqrt{2b_1 - 1}}{b_1^2} \tag{A7-8}$$

- Relative rotational velocity β_{2rx} of point m – From Eq. (1.5-3):

$$\beta_{2rx} = \sqrt{1 - \beta_{2tx}^2} \tag{A7-9}$$

- Increment of time Δt_m corresponding to increment z_m – From Eq. (1.4-18):

$$\Delta t_m = \frac{z_m r_0}{\beta_{2rx} c} = \frac{z_m}{\beta_{2rx}} \frac{2\pi r_0}{2\pi c} = \frac{z_m}{\beta_{2rx}} \frac{1}{2\pi f_0}$$

$$\Delta t_m = \frac{z_m}{\beta_{2rx}} \frac{1}{2\pi f_0} \tag{A7-10}$$

- Relative increment of time $\Delta t_m / T_2$ corresponding to z_m:

$$\frac{\Delta t_m}{T_2} = \frac{z_m}{\beta_{2rx}} \frac{1}{2\pi f_0} \frac{f_0}{b_1} = \frac{z_m}{\beta_{2rx}} \frac{1}{2\pi b_1}$$

$$\frac{\Delta t_m}{T_2} = \frac{z_m}{\beta_{2rx}} \frac{1}{2\pi b_1} \tag{A7-11}$$

- Maximum increment of relative translational velocity $\Delta\beta_{2tmax}$:

$$\Delta\beta_{2t\,\mathrm{max}} = \beta'_{2t} - \beta_{2t} = \frac{\sqrt{2b_1 - 1}}{b_1^2} - \frac{\sqrt{2b_1 - 1}}{b_1} = -\frac{(b_1 - 1)\sqrt{2b_1 - 1}}{b_1^2}$$

$$\Delta\beta_{2t\,\mathrm{max}} = \beta'_{2t} - \beta_{2t} = -\frac{(b_1 - 1)\sqrt{2b_1 - 1}}{b_1^2} \tag{A7-12}$$

- Instantaneous increments of relative translational velocity $\Delta\beta_{2t}$:

$$\Delta\beta_{2t} = \Delta\beta_{2t\,\mathrm{max}} \cos\left(\frac{2\pi f_0}{b_1} t\right) = -\frac{(b_1 - 1)\sqrt{2b_1 - 1}}{b_1^2} \cos\left(\frac{2\pi f_0}{b_1} t\right)$$

$$\Delta\beta_{2t} = -\frac{(b_1 - 1)\sqrt{2b_1 - 1}}{b_1^2} \cos\left(\frac{2\pi f_0}{b_1} t\right) \tag{A7-13}$$

Notes

Notes

<u>*Notes*</u>

REFERENCES

LIST OF PUBLICATIONS CONSULTED

A

Aczel, A.D., *The Mystery of the Aleph*, Washington Square Press, Published by Pocket Books, New York, London, Toronto, Sydney, 2001.

Aczel, A.D., *Entanglement*, A Plume Book, Penguin Group, New York, 2003.

Adair, R.K., *The Great Design – Particles, Fields and Creation*, Oxford University Press, New York, Oxford, 1987.

Agassi, J., *Faraday as a Natural Philosopher*, The University of Chicago Press, Chicago, IL, 1971.

Aiton, E.J., *The Vortex Theory of Planetary Motions*, American Elsevier, Inc., New York, 1972.

Aiton, E.J., *Leibniz - A Biography*, Adam Hilger Ltd, Bristol and Boston, 1985.

Akimov, A.E. and Shipov, G.I., "Torsion Fields and Their Experimental Manifestations," *Proc. of the Int'l Conference on New Ideas in Natural Sciences*, St. Petersburg, Russia, June 1996.

Akimov, A.E. and Tarasenko, V.Y., "Models of Polarized States of the Physical Vacuum and Torsion Fields," *Fizika*, No. 3, March 1992.

Albert, D.Z., "Bohm's Alternative to Quantum Mechanics," *Scientific American*, May 1994.

Albert, D.Z., *Quantum Mechanics and Experience*, Harvard University Press, Cambridge, Massachusetts, London, England, 1992,

Alexandersson, O., *Living Water - Victor Schauberger and the Secrets of Natural Energy*, Gateway Books, Bath, UK, 1996.

Alfven, H., *Worlds - Antiworlds: Antimatter in Cosmology*, W.H. Freeman and Company, San Francisco, 1966.

Allen, H.S., "The Case for a Ring Electron, *Proc. Phys. Soc. London*, vol. 31, pp 49-68 (1919).

Allen, R.E., *Greek Philosophy: Thales to Aristotle*, The Free Press, New York, 1966.

Allexander, A., *Infinitesimal*, Scientific American, Farrar, Straus and Giroux, New York, 2014.

Andrade, E.N. da C., *Rutherford and the Nature of the Atom*, Doubleday & Company, Inc., New York, 1964.

Andrulis, E.D., "Theory of the Origin, Evolution, and Nature of Life," (2012), *Life* **2012**, *2*, 1-30.

Arp, H., *Seeing Red - Redshifts, Cosmology and Academic Science*, Apeiron, Montreal, Canada, 1998.

Ash, D. and Hewitt, P., *The Vortex - Key to Future Science*, Gateway Books, Bath, England, 1991.

B

Babbitt, E.D., *The Principles of Light and Color*, Babbitt & Co., Kessinger Legacy Reprints, 1878.

Baggott, J., *The Meaning of Quantum Theory*, Oxford University Press, Oxford, UK, 1992.

Baker, J., *50 Physics Ideas You Really Need to Know,* Quercus, London, 2007.

Barrow, J.D. and Silk, J., *The Left Hand of Creation*, Oxford University Press, New York, 1983.

Barrow, J.D., *The Constants of Nature – The Numbers that Encode the Deepest Secrets of the Universe,* Vintage Books A Division of Random House, Inc., New York, 2002.

Barrow, J.D., *The Infinite Book – A Short Guide to the Boundless, Timeless and Endless*, Pantheon Books, New York, 2005.

Barrow, J.D., *New Theories of Everything – The Quest for Ultimate Explanation*, Oxford University Press Inc., New York, 2007.

Bartusiak, M., "Loops of Space," *Discover*, April 1993.

Bartusiak, M., "Gravity Wave Sky," *Discover*, July 1993.

Bastrukov, S.I. at al, "Spiral Magneto-Electron Waves in Interstellar Gas," Journal of Experimental and Theoretical Physics, Volume 93, pp 671-676, October 2001.

Beckmann, P., *A History of PI*, Dorset Press, New York, 1989.

Beiser, G., *The Story of Gravity - An Historical Approach to the Study of the Force That Holds the Universe Together*, E.P. Dutton & Co., Inc., New York, 1968.

Bekenstein, J.D., "Information in the Holographic Universe," *Scientific American*, pp. 59-65, August 2003.

Bentov, I., *Stalking the Wild Pendulum – On the Mechanics of Consciousness*, Density Book, Rochester, New York, 1977.

Bergman, D.L. and Wesley, J.P., "Spinning Charge Ring Model of Electron Yielding Anomalous Magnetic Moment," *Galilean Electrodynamics*, Vol. 1, No. 5, Sept./Oct. 1990.

Berke, J.P., Author, Editor, *Nanotubes and Nanowires (Selected Topics in Electronics and Systems*, World Scientific Publishing Company, Singapore, 2007.

Bernauer, J.C. and Paul, R., "The Proton Radius Problem," *Scientific American*, February 2014.

Biedermannn, H., *Dictionary of Symbolism –Cultural Icons and the Meaning behind Them*, A Meridian Book, New York, 1994.

Bhadkamkar, A. and Fox, H., "Electron Charge Cluster Sparking in Aqueous Solutions," *Journal of New Energy*, Vol. 1, No. 4, 1996.

Blackwood, O.H., et al, *An Outline of Atomic Physics*, John Wiley & Sons, Inc., New York, 1955.

Bloyd, J.G., *Broken Arrow of Time - Rethinking the Revolution in Modern Physics*, Writers Club Press, San Jose, CA, 2001.

Born, M., *Atomic Physics*, Dover Publications, Inc., Mineola, NY, 1989.

Boscovich, R.J., *A Theory of Natural Philosophy*, The M.I.T. Press, Cambridge, MA, 1966.

Boslough, J., *Stephen Hawking's Universe - An Introduction to the Most Remarkable Scientist of Our Time*, Quill/William Morrow, New York, 1985.

Boslough, J., *Masters of Time - Cosmology at the End of Innocence*, Addison-Wesley Publishing Company, Reading, MA, 1992.

Bostick, W., "Mass, Charge, and Current: The Essence of Morphology," *Physics Essays*, Vol. 4, No. 1, pp. 45-59, March 1991.

Bowers, B., *Michael Faraday and Electricity*, Priory Press Ltd., London, 1974.

Brennan, R.P., *Heisenberg Probably Slept Here – The Lives, Times and Ideas of the Great Physicists of the 20th Century*, John Wiley & Sons, Inc., New York, 1997.

Broglie, L., de, *The Revolution in Physics*, The Noonday Press, New York, 1953.

Broglie, L., de, *New Perspectives in Physics*, Basic Books, Inc. Publishers, New York, 1962.

Burger, T.J., *Nature* **271**, 402, 1978.

C

Calladine, C.R. and Drew, H.R., *Understanding DNA – The Molecule and How It Works*, Second Edition, Academic Press, New York, 2002.

Cambier, J-L., at al, "Theoretical Analysis of the Electron Spiral Toroidal Concept," NASA/CR-2000-210654, Dec. 2000.

Capra, F., *The Web of Life*, Anchor Books/Random House, Inc., 1997.

Capra, F., *The Tao of Physics*, Shambhala Publications, Inc., 1999.

Carter, J., *The Other Theory of Physics - A Non-Field Unified Theory of Matter and Motion*, Absolute Motion Press, 2000.

Carrigan, Jr., R.A. and Trower, W.P.. *Particles and Forces at the Heart of the Matter*, W.H. Freeman and Company, New York, 1990.

Carroll, R.L., *The Energy of Physical Creation,* The Carroll Research Institute, P.O. Box 3425, Columbia, S.C., 29230, 1985.

Carter, J., *The Other Theory of Physics – A Non-Field Unified Theory of Matter and Motion,* Absolute Motion Press, Enumclaw, Washington, 2000.

Cartledge, P., *Democritus*, Routledge, New York, 1999.

Caspar, M., *Kepler*, Abelard-Schuman, London and New York, 1992.

Cecil, T.E. and Chern, S., *Tight and Taut Submanifolds*, Cambridge University Press, New York, 1997.

Chalidze, V., *Mass and Electric Charge in the Vortex Theory of Matter*, Universal Publishers, 2001.

Chen, C., at al, "Equilibrium and Stability Properties of Self-Organized Electron Spiral Toroid," Physics of Plasma, Volume 8, Number 10, October 2001.

Chen, Y., Editor, *Nanotubes and Nanosheets: Functionalization and Applications of Boron Nitride and Other Nanomaterials,* CRC Press, London, 2015.

Clark, G., *The Man Who Tapped the Secrets of the Universe*, The University of Science and Philosophy, Swannanoa, Waynesboro, 2000.

Clawson, C.C., *Mathematical Sorcery – Revealing the Secrets of Numbers*, Perseus Books, Cambridge, MA, 2001.

Clawson, C.C., *Mathematical Mysteries – The Beauty and Magic of Numbers*, Perseus Publishing, Cambridge, MA, 1999.

Close, F., *Neutrino,* Oxford University Press, 2010.

Close, F., Marten, M., & Sutton, C., *The Particle Explosion*, Oxford University Press, New York, 1994.

Coats, C., *Living Energies*, Gateway Books, Bath, UK, 1996.

CODATA Recommended Values, *The NIST Reference on Constants, Units and Uncertainties*, 2011.

Consa, O., "Helical Model of the Electron," The General Science Journal, June 2014.

Cook, N., *The Hunt for Zero Point - Inside the Classified World of Antigravity Technology*, Broadway Books, New York, 2001.

Cook, T.A., *The Curves of Life*, Dover Publications, Inc., New York, 1979.

Collins, H. and Pinch T., *The Golem – What Everyone Should Know about Science,* Cambridge University Press, 1993.

Compton, A., "The Size and Shape of the Electron," *Phys. Rev. Second Series*, vol. 14, no.3, pp 247-259, (1919).

Coxeter, H.S.M., *Introduction to Geometry*, John Wiley & Sons, Inc., New York, 1961.

Coxeter, H.S.M., *The Beauty of Geometry*, Dover Publications, Inc., New York, 1968.

Crandall, B.C., *Nanotechnology – Molecular Speculations on Global Abundance,* The MIT Press, Cambridge, Massachusetts, London, England, 1996.

Crew, H., *The Wave Theory of Light - Memoirs by Huygens, Young and Fresnel*, American Book Company, New York, 1900.

Cushing, J.T., *Philosophical Concepts in Physics – The History Relations between Philosophy and Scientific Theories,* Cambridge University Press, 1998.

D

Dalton, J., et al, *Foundations of the Atomic Theory: Comprising Papers and Extracts*, Alembic Club, Edinburgh, UK, 1968.

Davies, P.C.W. and Brown, J., *Superstrings - A Theory of Everything?,* Cambridge University Press, Cambridge, UK, 1988.

Davies, P. and Gribbin, J., *The Matter Myth - Dramatic Discoveries That Challenge Our Understanding of Physical Reality*, Touchstone Book/Simon & Schuster, New York, 1992.

Davies, P., *About Time - Einstein's Unfinished Revolution*, Simon & Schuster, New York, 1995.

Day, W., *Bridge from Nowhere - The Photonic Origin of Matter*, Rhombics, Cambridge, MA, 1996.

Day, W., *A New Physics - Foundation for New Directions*, Cambridge, MA, 2000.

Derbyshire, J., *Unkn( )wn Quantity - A Real and Imaginary History of Algebra*, A Plume Book, Published by Penguin Group, New York, 2007.

Di Mario, D., "Electrogravity: A Basic Link Between Electricity and Gravity," *Speculations in Science and Technology*, Vol. 20, No. 4, Dec. 1997.

Dibner, B., *Oersted - And the Discovery of Electromagnetism*, Blaisdell Publishing Company, New York, 1962.

Dijksterhuis, E.J., *Archimedes*, Princeton University Press, Princeton, N.J., 1987.

Dirac, PAM, "Quantized Singularities in the Electromagnetic Fields," Scribd.com. 1931-05-29.

Dixon, R., *Mathographics*, Dover Publications, Inc, New York, 1991.

Dmitriyev, V.P., "Mechanical Analogy for the Wave - Particle: Helix on Vortex Filament," *Apeiron*, Vol. 8, No. 2, April 2001.

Domb, C., *Clerk Maxwell and Modern Science - Six Commemorative Lectures*, The Athlone Press, University of London, UK, 1963.

Drake, S., *Galileo at Work, His Scientific Biography*, The University of Chicago Press, Chicago, 1978.

Dresselhaus, M.S. and Eklund, P.C., *Science of Fullerences and Carbon Nanotubes: Their Properties and Applications*, Academic Press, 1996.

Drew, H.R., "The Electron as a Four-Dimensional Helix of Spin-1/2 Symmetry," *Physics Essays*, Vol. 12, No. 4, 1999.

Driscoll, R.B., *United Theory of Ether, Field and Matter*, Published by Author, Portland, OR, 1964.

Driscoll, R.B., *United Theory of Ether, Field and Matter (supplement)*, Published by Author, Oakland, CA, 1965.

Duncan, J.C., *Astronomy – A Textbook*, Fifth Edition, Harper & Brothers Publishers, New York, 1926.

E

Eckhart, L., *Four-Dimensional Space*, Indiana University Press, Bloomington, 1968.

Edwards, E.B., *Pattern and Design with Dynamic Symmetry*, Dover Publications, Inc., New York, 1932.

Edwards, L., *The Vortex of Life – Nature's Patterns in Space and Time*, Floris Books, Edinburgh, UK, 2006.

Ehrlich, R., *Crazy Ideas in Science - If You Might Even be True,* Princeton University Press, Prinston and Oxford, 2002.

Einstein, A. and Hopf, L., Ann. Phys., 33, 1096 (1910a): Ann. Phys., 33, 1105, 1910b.

Einstein, A., *Out of My Later Years*, A Citadel Press Book-Carol Publishing Group, New York, NY, 1991.

Einstein, A., *Relativity - The Special and the General Theory*, Crown Publishers, Inc., New York, 1961.

Einstein, A., Infeld, L., *The Evolution of Physics - From Early Concepts to Relativity and Quanta*, A Touchstone Book/Simon & Schuster, New York, 1966.

Einstein, A., "Aether and the Theory of Relativity," (Address on May 5, 1920, at the University of Leyden), *Journal of New Energy*, Vol. 7, No. 1, 2003.

Elgin, D., *The Living Universe – Where are We? Who are We? Where are We Going?*, Berrett-Koehler Publishers, Inc., San Francisco, CA, 2009.

Epstein, L.C., *Thinking Physics Is Gedanken Physics*, Insight Press, San Francisco, CA, 1983.

Epstein, L.C., *Relativity Visualized*, Insight Press, San Francisco, CA, 1992.

F

Farndon, J,. *The Great Scientists – From Euclid to Stephen Hawking,* Metro Books, New York, 2007.

Farrington, B., *Greek Science - Its Meaning for Us*, Penguin Books, Baltimore, MD, 1971.

Feber, A., "Supertwistors and Conformal Supersymmetry," *Nuclear Physics B* **132**: 55-64, 1978.

Ferguson, K., *Stephen Hawking - Quest for a Theory of Everything*, Bantam Books, New York, 1992.

Feynman, R.P., *Six Easy Pieces and Six Not-So-Easy Pieces,* Perseus Publishing, Cambridge, Massachusetts, 1995.

Flander, T.V., *Dark Matter, Missing Planets & New Comets – Paradoxes Resolved, Origins Illuminated,* Revised Edition, North Atlantic Books, Berkley, California, 1993.

Flood, R. and Lockwood, M., *The Nature of Time*, Basil Blackwell, Inc., Cambridge, MA, 1990.

Folger, T., "Tangled Up In Strings – Two Books Say That Today's Theoretical Physicists Are Way Off Course," *Discover*, Sept. 2006.

Ford, K.W., *101 Quantum Questions*, Harvard University Press, Cambridge, Massachusetts, 2011.

Fowler, P.W. and Manolopoulos, D.E., *An Atlas of Fullerenes*, Dover Publications, 2007.

Frank, P., *Einstein - His Life and Times*, Da Capo Press, Inc., New York, 1947.

Fraser, et al, *The Search for Infinity*, Facts on File, Inc., New York, 1995.

Freedman, D.H., "The Mysterious Middle of the Milky Way," *Discover*, November 1998.

Freeman, K. and McNamara, G., *In Search of Dark Matter*, Springer Praxis Publishing, Chichester, UK, 2006.

Friedman, N., *Bridging Science and Spirit - Common Elements in David Bohm's Rhysics, The Perennial Philosophy and Seth*, Living Lake Books, St. Louis, MO, 1994.

Fritzsch, H., *Quarks - The Stuff of Matter*, Basic Books, Inc., New York, 1983.

Fritzsch, H., *The Creation of Matter - The Universe From Beginning to End*, Basic Books, Inc., New York, 1984.

Funk & Wagnalls New Encyclopedia, Funk & Wagnalls, Inc., USA, 1966.

G

Gamow, G., *The Great Physicists from Galileo to Einstein,* Dover Publications, Inc., New York, 1988.

Gamow, G., *Thirty Years That Shook Physics - The Story of Quantum Theory*, Dover Publications, Inc., New York, 1985.

Gamow, G., *One, Two, Three ... Infinity - Facts and Speculations of Science*, Dover Publications, Inc., New York, 1988.

Gardner, M., *New Mathematical Diversions from Scientific American*, Simon and Schuster, New York, 1966.

Gardner, M., *Knotted Doughnuts and Other Mathematical Entertainments*, W.H. Freeman and Company, New York, 1986.

Gasperini, M., *The Universe Before the Big Bang,* Springer, Berlin, 2010.

Gauthier, R., "Faster-than-light quantum models of the photon and the electron", in M. S. El-Genk, (ed.) "*Space Technology and Applications International Forum – STAIF 2007*", American Institute of Physics 978-0-7354-0386-4/07, p1099-1108, 2007.

Gauthier, R., "Transluminal energy quantum models of the photon and the electron", in R.L. Amoroso, P. Rowlands & L.H. Kauffman (eds.) *The Physics of Reality: Space, Time, Matter, Cosmos, 8th Symposium in Honor of Mathematical Physicist Jean-Pierre Vigier*, Hackensack: World Scientific, 2013.

Gauthier, R., "A transluminal energy quantum model of the cosmic quantum", in R.L. Amoroso, P. Rowlands & L.H. Kauffman (eds.) *The Physics of Reality: Space, Time, Matter, Cosmos, 8th Symposium in Honor of Mathematical Physicist Jean-Pierre Vigier*, Hackensack: World Scientific, 2013.

Gautreau, R. and Savin, W., *Schaum's Outline of Theory and Problems of Modern Physics,* McGeaw-Hill, New York, 1978.

Gazale, M.J., *Gnomon*, Princeton University Press, Princeton, N.J., 1999.

Geerlings, G.K., *Wrought Iron In Architecture – An Illustrated Survey,* Dover Publications, Inc, New York, 1983.

Gell-Mann, M., *Quark and the Jaguar – Adventures in the Simple and the Complex*, A W, H. Freeman/Owl Book, Henry Holt and Company, LLC, New York, 1994.

Gell-Mann, M., *Complexity*, Vol. 1, no 5, John Wiley and Sons, Inc., New York, 1995/96.

Genz, H., *Nothingness - The Science of Empty Space*, Perseus Books, Reading, MA, 1999.

Geymonat, L., *Galileo Galilei: A Biography and Inquiry Into His Philosophy of Science*, McGraw-Hill Book Company, 1965.

Ghosh, A., *Origin of Inertia*, Apeiron, Montreal, Canada, 2000.

Ghyka, M., *The Geometry of Art and Life*, Dover Publications, Inc, New York, 1977.

Gillispie, C.C., *Dictionary of Scientific Biography*, Vol. IV, Charles Scribner's Sons, New York, 1971.

Ginzburg, V.L., *Theoretical Physics and Astrophysics*, Pergamon Press, 1979.

Ginzburg, V.L., *Physics and Astrophysics. A Selection of Key Problems*, Pergamon Press, 1985.

Gleick, J., *Chaos - Making a New Science*, Penguin Books, New York, 1988.

Gleick, J., *Genius - The Life and Science of Richard Feynman*, Pantheon Books, New York, 1992.

Gorini, C.A., *Geometry*, Facts On File, Inc., New York, 2003.

Goswami, A., *The Self-Aware Universe - How Consciousness Creates the Material World*, Penquin Putnam Inc., 1993.

Graver, J.E., "The Structure of Fullerene Signatures," *DIMACS Series in Discrete Mathematics and Theoretical Computer Science*, American Mathematical Society, 2005.

Gray, A., *Lord Kelvin - An Account of His Scientific Life and Work*, E.P. Dutton & Co., 1908.

Gray, A., *Modern Differential Geometry of Curves and Surfaces*, CRC Press, Boca Raton, 1993.

Greene, B., *The Elegant Universe - Superstrings, Hidden Dimensions, and the Quest for the Ultimate Theory*, W.W. Norton & Company, New York, 1999.

Greene, B., *The Fabric of the Cosmos*, Alfred A. Knopf, New York, 2004.

Gribbin, J., *Q Is For Quantum, An Encyclopedia of Particle Physics*, A Touchstone Book/ Simon & Schuster, New York, 2000.

Gribbin, J., *Schrödinger's Kittens and the Search for Reality,* Little, Brown & Company (Canada) Limited, 1995.

Guillen, M., *Five Equations that Changed the World*, Hyperion, New York, 1995.

H

Haisch, B., *The Purpose-Guided Universe – Believing in Einstein, Darwin, and God*, The Career Press, Inc., Franklin Lakakes, NJ, 2010.

Haisch, B., *The God Theory - Universes, Zero-Point Fields, and What's Behind It All*, Weser Books, San Francisco, CA, 2006..

Hall, A.R., *Isaac Newton - Adventurer in Thought*, Blackwell Publishers, Oxford, UK, 1994.

Hambidge, J., *Practical Applications of Dynamic Symmetry*, The Devin-Adair Company, New York, 1967.

Hargittai, I., Pickover, C.A., Editors, *Spiral Symmetry*, World Scientific Publishing Co., Singapore, 1992.

Harrington, P.S, *Star Watch*, John, Wiley & Sons, Inc., Hoboken New Jersey, 2003.

Harris, P.J.F., *Carbon Nanotube Science: Synthesis, Properties and Applications,* Cambridge University Press, Cambridge, UK, 2011.

Harrison, L.P., *Meteorology*, National Aeronautics Council, Inc., New York, 1942.

Hatch, E., *Modern Physics from a Classical Scale Perspective- Part 1 Concept Confirmed,* RWWAA Publication, Auburn, California, 2004.

Hawking, S., *Black Holes and Baby Universes and Other Essays*, Bantam Books, New York, 1993.

Hawking, S., *A Brief History of Time - From the Big Bang to Black Holes*, Bantam Books, New York, 1990.

Hawking, S., *On the Shoulders of Giants,* Running Press, Philadelphia, London, 2004.

Hawking, S. and Penrose, R., *The Nature of Space and Time*, Princeton University Press, Princeton, NJ, 2000.

Heath, J.L., *The Works of Archimedes*, Dover Publications, Inc., New York, 1953.

Heilbron, J.L., *The Dilemmas of an Upright Man - Max Planck as Spokesman for German Science*, University of California Press, Berkeley, 1986.

Heisenberg, E., *Inner Exile - Recollections of a Life with Werner Heisenberg*, Birkhauser Boston, MA, 1980.

Helmholtz, H., *On the Sensation of Tone – As a Physiological Basis for the Theory of Music,* Dover Publications, New York, 1954.

Henderson, L.D., *The Fourth Dimension and Non-Euclidean Geometry in Modern Art*, Princeton, 1983.

Herzberg, G., *Atomic Spectra and Atomic Structure*, Dover Publications, Inc., New York, 1944.

Hey, N., *Solar System*, Weidenfeld & Nicolson, The Orion Publishing Group, Wellington House, London, UK.

Hey, T. and Walters, P., *The New Quantum Universe*, Cambridge University Press, UK, 2003.

Hippel von, F., *Citizen Scientist*, A Touchstone Book/Simon & Schuster, New York, 1991.

Hoffmann, B., *Albert Einstein - Creator and Rebel*, New American Library, New York, 1972.

Hoffman, R.N., "Controlling Hurricanes – Can Hurricanes and Other Severe Tropical Storms be Moderated or Deflected?" *Scientific American*, Oct. 2004.

Hooft, G., *In Search of the Ultimate Building Blocks*, Cambridge University Press, UK, 1997.

Horgan, J., "Gravity Quantized? - A Radical Theory of Gravity Weaves Space From Tiny Loops," *Scientific American*, September 1992.

Hotson, D.L., "Dirac's Equation and the Sea of Negative Energy, Part 1," *Infinite Energy*, Vol. 8, Issue 43, 2002.

Hotson, D.L., "Dirac's Equation and the Sea of Negative Energy, Part 2," *Infinite Energy*, Vol. 8, Issue 44, 2002.

Hurley, W.M., *Prehistoric Cordage – Aldine Manuals on Archeology 3,* Taraxacum, Washington, 1979.

I

Icke, V., "From Expansion to Intelligence in the Universe," *Speculations in Science and Technology*, Vol. 14, No. 4, 1991.

Ipsen, D.C., *Archimedes: Greatest Scientist of the Ancient World*, Enslow Publishers, Inc., Hillside, N.J., 1988.

J

Jammer, M., *Concepts of Mass in Classical and Modern Physics*, Dover Publications, Inc., Mineola, New York, 1997.

Jammer, M., *Concepts of Force*, Dover Publications, Inc., Mineola, New York, 1999.

Jean, Sir, J, *Science & Music,* Dover Publications, Inc., New York, 1968

Johnson, G., "The Inelegant Universe – Two New Books Argue That It Is Time For String Theory To Give Way," *Scientific American*, September, 2006.

Jefimenko, O.D., *Gravitation and Cogravitation*, Electret Scientific Company, Star City, West Virginia, 2006.

Jones, B.Z., *The Golden Age of Science,* Simon and Schuster, New York, 1966.

Jonsson, I., *Emanuel Swedenborg*, Twayne Publishers Inc., New York, 1971.

K

Kafatos, M. and Nadeau, R., *The Conscious Universe - Part and Whole in Modern Physical Theory*, Springer-Verlag New York, Inc., New York, 1990.

Kaku, M., *Physics of the Impossible,* Anchor Book, A Division of Random House, Inc., 2008.

Kaku, M., *Visions – How Science Will Revolutionize the 21st Century,* Anchor Books, Doubleday, New York, London, 1997.

Kaku, M., *Beyond Einstein - The Cosmic Quest for Theory of the Universe*, Anchor Books/Doubleday, New York, 1995.

Kaku, M., *Hyperspace*, Anchor Books/Doubleday, New York, 1994.

Kaku, M., *Introduction to Superstrings*, Springer-Verlag, New York, 1988.

Kanarev, F.M., "Model of the Electron," APERON, Vol. 7. Nr. 3-4, July-October, 2000.

Kanarev, F.M., *The Foundation of Physchemistry of Microworld*, Kuban State Agrarian University (KSAU), Krasnodar, Russia, 2002.

Kane, G., *The Particle Garden - Our Universe as Understood by Particle Physicists*, Addison-Wesley Publishing Company, Reading, MA, 1995.

Kanigel, R., *The Man Who Knew Infinity - A life of the Genius Ramanujan,* Washington Square Press, Published by Pocket Books, New York, London, 1991.

Kaplan, R., *The Nothing That Is - A Natural History of Zero*, Oxford University Press, Oxford, New York, 1999.

Kaplan, R. and Kaplan, E., *The Art of The Infinite*, Oxford University Press, Oxford, New York, 2003.

Kaufmann, W.J., *Black Holes and Warped Spacetime*, W.H. Freeman and Company, San Francisco, CA.

Kimura, Y.G., *The Book of Balance*, (Translation), The University of Science and Philosophy, Contact Printing, North Vancouver, B.C., Canada, 2002.

Kimura, Y.G., "The Transcendent Unity of Science and Spirituality," *VIA – Vision in Action*, Vol. 2, No. 1 & 2, 2004.

King, M.B., "Vortex Filaments, Torsional Fields and the Zero-Point Energy," *Journal of New Energy*, Vol. 3, No. 2/3, 1998.

King, M.B., "Dual Vortex Forms: The Key to a Large Zero-Point Energy Coherence," *Journal of New Energy*, Vol. 5, No. 2, 2000.

King, M.B., *Quest for Zero Point Energy*, Adventures Unlimited Press, Kempton, IL, 2001.

King, M.B., *Tapping the Zero-Point Energy,* Paraclete Publishing, Provo, Utah, 1989.

Knight, D.C., *The Science Book of Meteorology*, Franklin Watts, Inc., New York, 1964.

Krauss, L.M., *Quintessence - The Mystery of Missing Mass in the Universe*, Basic Books, New York, NY, 2000.

Kumar, S., "A Spiral Structure for Elementary Particles," Int. J. Res. Vol. 1, Issue 6, July 2014.

Kumar, S., "Journey of the Universe from Birth to Rebirth with Insight into Unified Interaction of Elementary Particles with Spiral Structure," Int. J. Res. Vol. 1, Issue 9, October 2014.

Kumar, S., "Quantum Spiral Theory," Int. J. Res. Vol. 2, Issue 1, January 2015.

Kumar, S., "Spiral Hashed Information Vessel," International Journal of Scientific & Engineering Research, Vol. 4, Issue 6, April 2015.

Kumar, S., "Spiral Structure of Elementary Particles Analogous to Sea Shells: A Mathematical Description," International Journal of Current Research, Vol. 7, Issue 02, Feb. 2015, p. 12814.

Kumar, S., "Mass-Energy Equivalence in Spiral Structure for Elementary Particles and Balance of Potentials," International Journal of Scientific & Engineering Research. Vol. 6, Issue 6, July 2015.

L

Lamb, G.L., "Solutions and the Motion of Helical Curves," *Physical Review Letters*, Vol. 37, No. 5, August 1976.

Lakhtakia, A. and Weiglhofer, W.S., "Time-Dependent Beltrami Fields in Free Space: Dyadic Green Functions and Radiation Potentials," *Physical Review E*, Vol. 49, Number 6, June 1994.

Lakhtakia, A. and Weiglhofer, W.S., "Covariances and Invariances of the Beltrami-Maxwell Postulates," *IEE Proc. - Sci. Meas. Technol.*, Vol. 142, No. 3, May 1995.

Lang, T.G., "Proposed Unified Field Theory – Part II: Protons, Neutrons and Fields," *Galilean Electrodynamics*, Vol. 12, No. 6, Nov./Dec. 2001.

Larsen, R., et al, *Emanuel Swedenborg - A Continuing Vision*, Swedenborg Foundation, Inc., New York, 1988.

Laugwitz, D., *Differential and Riemannian Geometry*, Academic Press, New York, 1965.

Lauwerier, H., *Fractals - Endlessly Repeated Geometrical Figures*, Princeton University Press, Princeton, New Jersey, 1991.

Lederman, L.M. and Teresi, D., *The God Particle*, Bantam Doubleday Dell Publishing Group, Inc., New York, 1993.

Lederman, L. and Hill, C.T., *Symmetry and the Beautiful Universe*, Prometheus Books, New York, 2004.

Lederman, L.M. and Hill, C.T., *Quantum Physics for Poets*, Prometheus Books, New York, 2011.

Lerner, E.J., *The Big Bang Never Happened*, Vintage Books, Random House, Inc., New York, 1992.

Lewis, H., *Geometry – A Contemporary Course,* Third Edition, McCormick-Mathers Publishing Company, Cincinnati, Ohio, 1973.

Lewis, J.R., *Scientology*, Cary, NC, Oxford University Press, 2009.

Lindgren, C.E., *Four-Dimensional Descriptive Geometry*, McGraw-Hill Book Company, New York, 1968.

Lindley, D., *The End of Physics - The Myth of a United Theory*, HarperCollins Publishers, Inc., 1993.

Lipschultz, M.M, *Differential Geometry,* Schaum's Outline Series, McGraw-Hill, New York, 1969.

Livio, M., *The Equation That Couldn't Be Solved – How Mathematical Genius Discovered the Language of Symmetry*, Simon and Schuster, New York, 2005.

Livio, M., *The Golden Ratio*, Broadway Books, New York, 2002.

Lockwood, E.H., *A Book of Curves*, Cambridge University Press, New York, 1961.

Lomberg, J., *Unified Force Theory, Dark Matter and Consciousness,* The Acnor Trust, PO Box 4706, Salem, Oregon, 2004.

Lorentz, H.A., *Problems of Modern Physics - A Course of Lectures Delivered in the CA Institute of Technology*, Ginn and Company, Boston, 1927.

Lucas, C.W., "A Classical Electromagnetic Theory of Elementary Particles," *Journal of New Energy*, Vol. 6, No. 4, 2002.

Lucas, C.W. "A Classical Electromagnetic Theory of Elementary Particles Part 2, Interwining Charge-Fibers," The Journal of Common Sense Science, Foundation of Science, May 2005, Vol. 8 No. 2.

Ludwig, C., *Michael Faraday - Father of Electronics*, Herald Press, Scottdale, PA, 1978.

Lugt, H.J., *Vortex Flow in Nature and Technology*, Krieger Publishing Company, Malabar, Florida, 1995.

Lykken, J. and Spiropulu, M., "Supersymmetry and the Crisis in Physics," *Scientific American*, May 2014.

M

Maldacena, J., "The Illusion of Gravity," *Scientific American*, pp. 57-63, November 2005.

Magueijo, J., *Faster Than the Speed of Light - The Story of A Scientific Speculation,* Penguin Books, New York, 2003.

Manning, J., *The Coming Energy Revolution - The Search for Free Energy*, Avery Publishing Group, Garden City Park, New York, 1996.

Manning, H.P., *The Fourth Dimension Simply Explained*, Dover Publications, Inc., New York, 1960.

Maor, E., *e: The Story of a Number*, Princeton University Press, Princeton, NJ, 1994.

Maor, E., *Trigonometric Delights*, Princeton University Press, Princeton, NJ, 1998.

Marsden, J.E. and McCracken, M., *The Hopf Bifurcation and Its Applications*, Springler-Verlag, New York, Berlin, 1976.

Mazur, B., *Imagining Numbers (particularly the square root of minus fifteen)*, Picador, New York, 2003.

McCrea, W.H., "Arthur Stanley Eddington," *Scientific American*, June 1991.

McCutcheon, M., *The Final Theory – Rethinking Our Scientific Legacy*, Universal Publishers, Boca Raton, Florida, 2004.

McLeish, J., *The Story of Numbers*, Fawcett Columbine, New York, 1991.

Meacher, M., *Destination of the Species – The Riddle of Human Existence*, Books, Winchester, UK, Wasington USA, 2009.

Melker, A.A. and Krupina, M.A., "Designing Muni-Fullerences and Their Relatives on Graph Basis," *Materials Physics and Mechanics 20*, 18-24 2014.

Messent, J., *Embroidery & Architecture*, B.T Batsford Ltd., London, 1985.

Millar, D., et al, The Cambridge Dictionary of Scientists, Cambridge University Press, 1996.

Miller, A.I., *137 – Jung, Pauli and the Pursuit of the Scientific Obsession*, W.W Norton & Company, Inc., New York, 2009.

Miller, A.I., *Albert Einstein's Special Theory of Relativity,* Springer-Verlag, New York, Berlin, 1997.

Milton, R., *Alternative Science - Challenging the Myths of the Scientific Establishment*, Park Street Press, Rochester, Vermont, 1996.

Mitchell, W.C., *Bye Bye Bing Bang – Hello Reality,* Cosmic Sense Books, Carson City, Nevada, 2002.

Mitsopoulos, T.D., "Similarity Between Elementary Particles and Electric Circuits," *Galilean Electrodynamics*, Vol. 12, No. 6, Nov./Dec. 2001.

Moore, W., *Schrodinger - Life and Thought*, Cambridge University Press, UK, 1992.

Mortimer, S., *Techniques of Spiral Work – A Practical Guide to the Craft of Making Twists by Hand,* Linden Publishing, Fresno, California, 1995.

Moyer, M., "Is the Space Digital," *Scientific American*, February 2012.

Mugnai, D., et al, "Observation of Superluminal Behaviors in Wave Propagation," *Physical Review Letters*, Vol. 84, Number 21, May 2000.

Murchie, G., *The Seven Mysteries of Life - An Exploration in Science and Philosophy*, Houghton Mifflin Company, Boston, 1978.

N

Nahin, P.J., *An Imaginary Tale – The Story of $\sqrt{-1}$*, Princeton University Press, Princeton, New Jersey, 1998.

Nakahara, M., *Geometry, Topology and Physics,* Second Edition, Taylor & Francis, Taylor & Francis Group, New York, London, 2003.

Nash, C. and Sen, S., *Topology and Geometry for Physicists*, IBI Global,

London, UK, 1983.

Nernst, W., Verh. Dtsch. Phys. Ges., 18, 83, 1916.

Newton, I., *The Principia*, Prometheus Books, Amherst, New York, 1995.

Nierengarten, J-F., Editor, *Fullerenes and Other Carbon-Rich Nanostructures (Structure and Bonding)*, Springer, 2014.

Niven, W.D., *The Scientific Papers of James Clerk Maxwell*, Dover Publications, Inc., New York, 1890.

Novikov, I.D., *The River of Time*, Cambridge University Press, Cambridge, UK, 1998.

O

Okun, L.B., "The Concept of Mass," *Physics Today*, Vol. 42, June 1989.

Oliwensrein, L., "Bent out of Shape," *Discover*, July 1993.

Oros di Bartini, R., "Relations Between Physical Constants," *Progress in Physics*, v. 3, pp. 34-40, October 2005.

Oros di Bartini, R., "Some Relations Between Physical Constants," *Doklady* Acad. Nauk USSR, v. 163, No. 4, pp. 861-864, 1965.

Oschman, J.L. and Schman N.H., "Vortical Structure of Light and Space: Biological Implications," *J Vortex Sci Technol.* 2:1, 2015.

Oschman, J.L. and Schman N.H., "The Heart as a Bi-Directional Scalar Field Antena," *J Vortex Sci Technol.* 2:2, 2015.

Oschman, J.L., *Energy Medicine – The Scientific Basis,* Second Edition, Elsevier, New York, 2016.

P

Pagels, H.R., *The Cosmic Code – Quantum Physics as the Language of Nature*, Bantam Books, New York, 1982.

Panek, R., *The 4% Universe – Dark Matter, Dark Energy, and the Race to Discover the Rest of Reality*, Houghton Mifflin Harcourt, Boston, New York, 2011.

Pappas, T., *The Joy of Mathematics – Discovering Mathematics All Around You,* Wide World Publishing, Tetra, 1989.

Parry, A., *The Russian Scientist*, The Macmillan Company, New York, 1973.

Parson, A.L., "A Magneton Theory of the Structure of the Atom," Smithsonian Miscellaneous Collections, Vol. 65, No. 11, Publication 2371, Nov. 29, 1915.

Pauli, W., *Theory of Relativity*, Pergamon Press, London, UK, 1958.

Peat, F.D., *Superstrings and the Search for The Theory of Everything*, Contemporary Books, Lincolnwood (Chicago), ILL,1988.

Pedoe, D., *Geometry - A Comprehensive Course*, Dover Publications, Inc., New York, 1988.

Peebles, P.J.E., *Principles of Physical Cosmology*, Princeton University Press, Princeton, New Jersey, 1993.

Penrose, R., "Twistor Quantization and Curved Space-time." *International Journal of Theoretical Physics* (Springer Netherlands), **1**: 61-99, 1968.

Penrose, R., *The Road to Reality - A Complete Guide to the Laws of the Universe*, Alfred A. Knopf, New York, 2005.

Penrose, R., *Shadows of the Mind – A Search for the Missing Science of Consciousness,* Oxford University Press, 1994.

Peratt, A.L., "Birkeland and the Electromagnetic Cosmology," *Sky & Telescope*, May 1985.

Physical Review D: Particles and Fields, Vol. 54, The American Physical Society, 1996.

Pierson, H.O., *Handbook of Carbon, Graphite, Diamond and Fullerenes: Properties and Applications (Material Science and Process Technology)*, Noyes Publications, 1994.

Pickover, C.A., *Mathematics and Beauty II; Spirals and "Strange" Spirals in Civilization, Nature, Science, and Art*, IBM Thomas J. Watson Research Center, Yorktown Heights, NY, 1987.

Polyakov, A., "Gauge Fields and Strings," Harwood Academic Publishers 1987, *Nucl. Phys.* **B396**, 367, 1993.

Ponomarev, C.D. and Andreeva, L.E., *The Calculation of Elastic Elements of Machines and Sensors*, Machinostroenie, Moscow, 1980.

Porter, R., *The Biographical Dictionary of Scientists*, Oxford University Press, New York, 1994.

Posamentier, A.S. and Lehmann, I., *A Biography of the World's Most Mysterious Number*, Prometheus Books, Amherst, New York, 2004.

Potemra, T. A., "Hannes Alfven, Father of Space Plasma Physics," Geomagne-tism and Aeronomy with Special Historical Case Studies, IAGA   Newsletters 29/1997, Published by IAGA, Germany, p.101, 1997.

Price, W.C., et al, *Wave Mechanics; The First Fifty Years - A Tribute to Professor Louis De Broglie*, John Wiley & Sons, New York-Toronto, 1973.

Price, H., *Time's Arrow and Archimedes' Point*, Oxford University Press, New York, 1996.

Purce, J., *The Mystical Spiral- Journey of the Soul,* Thames and Hudson, 1974.

Purdy, S., and Sandak, C.R., *Ancient Greece*, Franklin Watts, New York, 1982.

Puthoff, H.E., et al, "Engineering the Zero-Point Field and Polarizable Vacuum for Interstellar Flight," *Journal of New Energy*, Vol. 6, No. 1, 2001.

R

Randless, J. *Breaking the Time Barrier*, Paraview Pocket Books, New York, London, 2005,

Reed, D., "Excitation and Extraction of Vacuum Energy Via EM-Torsional Field Coupling - Theoretical Model," *Journal of New Energy*, Vol. 3, No. 2/3, 1998.

Reed, D., "A New Paradigm for Time – Evidence From Empirical and Esoteric Sources," *Journal of New Energy*, Vol. 6, No. 2, 2001.

Resnick, R., *Introduction to Special Relativity,* Jon Wiley & Sons, Inc., New York, London, 1968.

Ridley, B.K., *Time, Space and Things*, Cambridge University Press, Cambridge, UK, 1994.

Riordan, M., *The Hunting of the Quark - A True Story of Modern Physics*, Simon and Schuster/Touchstone, New York, 1987.

Riordan, M. and Schramm, D.N., *The Shadows of Creation - Dark Matter and the Structure of the Universe*, W.H. Freeman and Company, New York, 1991.

Rucker, R., *The Fourth Dimension*, Houghton Mifflin Company, Boston, 1984.

Russell, P., *The White Hole in Time*, Harper San Francisco, 1992.

Russell, W. *The Universal One*, University of Science and Philosophy, Swannanoa, Waynesboro, Virginia, 1974.

Russell, W., *A New Concept of the Universe*, The University of Science and Philosophy, Swannanoa, Waynesboro, VA, 1989.

Russell, W., *The Secret of Light*, The University of Science and Philosophy, Swannanoa, Waynesboro, VA, 1994.

Ryu, C., *The Grand Unified Theory – A Scientific Theory of Everything*, PublishAmerica, Baltimore, 2004.

S

Saito, R., Author, Editor, *Physical Properties of Carbon Nanotubes*, Imperial College Press, London, 1998.

Salem, K.G., *The New Gravity - A New Force - A New Mass - A New Acceleration - Unifying Gravity with Light*, Salem Books, Johnstown, PA, 1994.

Sagan, C., *Cosmos,* Ballantine Books, New York, 1980.

Sanders, P.A. Jr., *Scientific Vortex Information*, Free Soul Publishing, Sedona, AZ, 1992.

Sano, C., "Twisting & Untwisting of Spirals of Aether and Fractal Vortices Connecting Dynamic Aethers," *Journal of New Energy*, Vol. 6, No. 2, 2001.

Sarg, S., "A Physical Model of the Electron According to the Basic Structure of Matter Hypothesis," *Physics Essays*, Vol. 16, No.2, 180-195, 2003.

Sarg, S, "Basic Structure of Matter – Supergravitation Unified Theory Based on an Alternative Concept of the Physical Vacuum," Proceedings of the 17th Annual Conference of the NPA at Long Beach, CA, Vol. 7, pp. 479-484, 23-26 June, 2010.

Savov, E., *Theory of Interaction - The Simplest Explanation of Everything*, Geones Books, Sofia, Bulgaria, 2002.

Schneider, M.S., *A Beginner's Guide to Constructing the Universe*, HarperPerennial, New York, 1995.

Schweighauser, C.A., *Astronomy from A to Z – A Dictionary of Celestial Objects and Ideas,* Sangamon State University, Springfield, Illinois, 1991.

Schwenk, T., *Sensitive Chaos – The Creation of Flowing Forms in Water and Air,* Rudolf Steiner Press, Hillside House, East Sussex, 2008.

Schwerdtfeger, H., *Geometry of Complex Numbers – Circle Geometry, Moebius Transformation, Non-Euclidean Geometry*, Dover Publication, Inc., New York, 1979.

Scientific American, *The Enigma of Weather*, Scientific American, New York, NY, Oct. 2004.

Segal, V.M., "Materials Processing by Simple Shear," *Mat. Sci & Eng.,* vol. 197, 157-164, 1995.

Segal, V.M., "Equal Channel Angular Extrusion: From Macro Mechanics to Structure Formation," *Mat. Sci & Eng.,* vol. 271, 322-333, 1999.

Segal, V.M., "Severe Plastic Deformation: Simple Shear versus Pure Shear," vol. 338, pp. 331-344, 2002.

Seggern, D.H. von, *CRC Handbook of Mathematical Curves and Surfaces*, CRC Press, Boca Raton, Florida, 1990.

Seggern, D.H. von, *CRC Standard Curves and Surfaces*, CRC Press, Boca Raton, Florida, 1993.

Segre, E., *Nuclei and Particles - An Introduction to Nuclear and Subnuclear Physics*, W.A. Benjamin, Inc., New York, 1965.

Seife, C., *Zero - The Biography of a Dangerous Idea*, Viking Penquin, New York, 2000.

Semat, H., *Introduction to Atomic and Nuclear Physics,* Rinehart & Company, Inc., New York, 1958.

Series, G.W., *Advances - The Spectrum of Atomic Hydrogen*, World Scientific, New Jersey, 1988.

Serway, R.A., *Physics for Scientist & Engineers with Modern Physics*, 3rd Edition, Saunders Golden Sunburst Series, Saunders College Publishing, Philadelphia, PA, 1990.

Seward, C., "Ball Lightning Events Explained as Self-Stable Spinning High-Density Plasma Toroids or Atmospheric Spheromaks," IEEE *Access* Practical Innovations, Volume 2, 2014, 153-59.

Sharlin, H.I., *Lord Kelvin - The Dynamic Victorian*, The Pennsylvania State University Press, PA, 1979.

Sheka, E., *Fullerences: Nanochemistry. Nanomagnetism, Nanomedicine, Nanophotonics,* CRC Press, 2011.

Siegfried, T., *Strange Matters – Undiscovered Ideas at the Frontiers of Space and Time,* The Berkley Publishing Group, A division of Penguin Group, New York, 2004.

Siegfried, T., *The Bit and the Pendulum*, John Wiley & Sons, Inc., New York, 2000.

Simhony, M., *Invitation to the Natural Physics of Matter, Space, Radiation*, World Scientific, New Jersey, London, 1994.

Smolin, L., *The Trouble With Physics: The Rise of String Theory, The Fall of a Science, and What Comes Next,* Houghton Mifflin, 2006.

Sprott, J.C., *Strange Attractions - Creating Patterns in Chaos*, M&T Books, New York, 1993.

Sproull, R.L., *Modern Physics – A Textbook for Engineers,* John Wiley & Sons, New York, 1956.

Stenger, V.J., *God and the Atom – From Democritus to the Higgs Boson: The Story of a Triumphant Idea*, Prometheus Books, New York, 2013.

Sternberg, S., *Curvature in Mathematics and Physics*, Dover Publications, Inc., New York, 2012.

Sternglass, E.J., *Before the Big Bang - The Origins of the Universe*, Four Walls Eight Windows, New York, NY, 1997.

Strogatz, S., *Sync - The Emerging Science of Spontaneous Order*, Hyperion, New York, 2003.

Sunden, O., "Time-Space-Oscillation: Hidden Mechanism Behind Physics," *Galilean Electrodynamics*, Vol. 12, Special Issue 2, Fall 2001.

Swedenborg, E., *The Principia*, Swedenborg Society, London, 1912.

Synge, J.L., "The Electrodynamic Double Helix," In *Magic Without Magic: John Archibald Wheeler* - A Collection of Essays in Honor of His Sixtieth Birthday, edited by John R. Klauder, W. H. and Company, San Francisco, 1972.

T

Tanaka, K., Editor, Iijima, S., *Carbon Nanotubes and Graphene, Second Edition,* Nanotube Research Center, Tsukuba, Japan, 2014.

Talbot, M., *The Holographic Universe*, Harper Perennial, 1992.

Tewari, P., *Universal Principles of Spacetime and Matter - A Call for Conceptual Reorientation,* Crest Publishing House, New Deli, 2002.

Tewari, P., "On the Space-Vortex Structure of the Electron," www.tewari.org/ Theory_Papers/Tewari-Final%20Proof.pdf. 2005.

Thomson, J.J., *A Treatise on the Motion of Vortex Rings*, MacMillan and Co., London, 1883.

Thomson, J.J., *Electricity and Matter*, Charles Scribner's Sons, New York, 1904.

Thomson, D.W. and Bourassa, J.D., *Secrets of Aether*, Published by The Aenor Trust, Salem, OR, 2004.

Time-Life Books, *A Soaring Spirit - Time Frame BC 600-400*, The Time Inc. Book Company, Alexandria, VA, 1987.

Time-Life Books, *Empires Ascendant - Time Frame 600 BC - AD 200*, The Time Inc. Book Company, Alexandria, VA, 1987.

Tricker, R.A., *The Contributions of Faraday and Maxwell to Electrical Science*, Pergamon Press Ltd., London, UK, New York, 1987.

Thorne, K.S., *Black Holes & Time Warps - Einstein's Outrageous Legacy*, W.W. Norton & Company, New York, 1994.

Treasures of Early Irish Art: 1500 B.C. to 1500 A.D., From the Collections of the National Museum of Ireland, Royal Irish Academy, Trinity College, Dublin, 1977.

U

Unger, R.M. and Smolin, L., *The Singular Universe and the Reality of Time*, Cambridge University Press, Cambridge, UK, 2015.

V

Von Stade, S., "Owner". *Flowtoys.* Flowtoys (Toroflux).

Valens, E.G., *The Attractive Universe: Gravity and Shape of Space*, Motion, Magnet, 1969.

Van der Laan, C., "The Vortex Theory of Atoms," Thesis for the Master's Degree in History and Philosophy of Science, Institute for History and Foundation of Science, Utrecht University, Dec. 2012.

Van Eenwik, J.R., *Archetypes & Strange Attractors - The Chaotic World of Symbols*, Inner City Books, Toronto, Canada, 1997.

Van Flandern, T., *Dark Matter, Missing Planets & New Comets - Paradoxes Resolved, Origins Illuminated*, North Atlantic Books, Berkeley, CA, 1993.

Valone, T., "Inside Zero Point Energy," *Journal of New Energy*, Vol. 5, No. 4, 2001.

Van Nostrand's Scientific Encyclopedia, D. Van Nostrand Company, Inc., 1958.

Veltman, M., *Facts and Mysteries in Elementary Particle Physics*, World Scientific, New Jersey, 2003.

Venable, W.M., *The Interpretation of Spectra*, Reinhold Publishing Corporation, New York, 1948.

Volk, G., "Toroids, Vortices, Knots, Topology and Quanta," Proceedings of of the 18th Annual Conference the NPA, 6-9 at the University Maryland College Park, MD, Vol. 8, July 2011.

Vrooman, J.R., *Rene Descartes - A Biography*, G.P. Putnam's Sons, New York, 1970.

W

Wagner, O.E., "Structure in the Vacuum," *Frontier Perspectives*, Vol. 10, No. 2, Fall 2001.

Walker, F.L., "The Fluid Space Vortex: Universal Prime Mover," *Physics Essays*, Vol. 15, No. 2, 2002.

Wallace, D.F., *Everything and More - A Compact History of ∞,* W.W. Norton & Company, New York, London, 2003.

Watson, J.D., *The Double Helix*, W.W. Norton & Company, New York, 1980.

Weber, C.S., "VRML Gallery of Fullerenes," Fullerene Library, JSV1.08, 1999.

Weinberg, S., *Dreams of a Final Theory*, Pantheon Books, New York, 1992.

Weir, S.T., Mitchell, A.C. and Nellis, W.J., "Metallization of Fluid Molecular Hydrogen," *Physics Review Letters* 76, 1860, 1996.

Westfall, R., *The Life of Isaac Newton*, Cambridge University Press, New York, NY, 1994.

Wheeler, J.A., *Geons, Black Holes & Quantum Foam – A Life in Physics,* W.W. Norton & Company, Inc., New York, 1998.

Wheeler, J.A., *Geometrodynamics, Topics of Modern Physics, Vol.1*, Academic Press Inc., New York, NY, 1962.

White, H.E., *Introduction to Atomic Spectra*, McGraw-Hill Book Company, Inc., New York, 1934.

Whitney, S.K., "9 Editor's Essays," *Galilean Electrodynamics*, Vol. 16, Special Issue 3, Winter 2005.

Whitney, S.K., *Algebraic Chemistry – Applications and Origins,* Nova Science Publishers, Inc., New York, 2013.

Wiener, N., *Cybernetics or Control and Communication in the Animal and the Machine*, 2nd edition, The MIT Press and John Wiley & Sons, Inc., New York, 1961.

Wigner, E. and Huntington, H.B., "On the Possibility of a Metallic Modification of Hydrogen, *Journal of Chemical Physics* **3** (12): 764, 1935

Wilczek, F., *Longing for the Harmonies – Themes and Variations from Modern Physics*, W.W. Norton & Company, New York, 1987.

Wilczek, F., *The Lightness of Being – Mass, Ether, and the Unification of Forces*, Basic Books, New York, 2008.

Wilczek, F., *A Beautiful Question – Finding Nature's Deep Design*, Penguin Press, New York, 2015.

Williamson, J.G, and van der Mark, M.B., "Is the electron a Photon with Toroidal Topology?," Annales de la Fondation Louis de Broglie, Volume 22, No. 2, 133, (1997).

Witten, E., "Perturbative Gauge Theory as a String Theory in Twistor Space," (2004) (http://arxiv.org/abs/hep-th/0312171)" *Commun Math. Phys.* 252: 189-258.

Woit, P., *Not Even Wrong: The Failure of String Theory and the Search for Unity in Physical Law*, Basic Books, 2006.

Wolff, M., "Origin of the Mysterious Instantaneous Transmission of Events in Science," *The Cosmic Light*, Vol. 4, No. 2, Spring 2002.

Wolfram, S., *A New Kind of Science*, Wolfram Media, LLC, Champaign, IL, 2002.

Wong, H.-S. P., *Carbon Nanotubes and Graphene Device,* Cambridge University Press, Cambridge, UK, 2011.

Z

Zeman, R.K. et al., *Helical/Spiral CT - A Practical Approach*, McGraw-Hill, Inc., New York, 1995.

Zombeck, M.V., *Handbook of Space Astronomy and Astrophysics*, Cambridge University Press, Cambridge, UK, 1990.

Zwikker, C., The Advanced Geometry of Plane Curves and Their Applications, Dover Publications, Inc., New York, 1994.

AUTHOR'S PUBLICATIONS RELATED TO THIS BOOK

1. BOOKS

1. Ginzburg, V.B., *Spiral Grain of the Universe - In Search of the Archimedes File*, University Editions, Inc., Huntington, WV, 1996.

2. Ginzburg, V.B., *Unified Spiral Field and Matter - A Story of a Great Discovery*, Helicola Press, Pittsburgh, PA, 1999.

3. Ginzburg, V.B., *The Unified Spiral Nature of the Quantum & Relativistic Universe*, First edition, Helicola Press, Pittsburgh, PA, 2002.

4. Ginzburg, V.B., *The Unification of Strong, Gravitational & Electric Forces*, Helicola Press, Pittsburgh, PA, 2003.

5. Ginzburg, V.B., *Prime Elements of Ordinary Matter, Dark Matter & Dark Energy*, Helicola Press, Pittsburgh, PA, 2006.

6. Ginzburg, V.B., *Prime Elements of Ordinary Matter, Dark Matter & Dark Energy – Beyond Standard Model & String Theory*, Second revised edition, Universal Publishers, Boca Raton, Florida, 2007.

7. Ginzburg, V.B., *The Spacetime Origin of the Universe*, First edition, Helicola Press, Pittsburgh, PA, 2013.

8. Ginzburg, V.B., *The Spacetime Origin of the Universe with Visible Dark Matter & Energy*, Third edition, Helicola Press, Pittsburgh, PA, 2016.

2. ARTICLES & PATENTS

1. Ginzburg, V.B., "Toroidal Spiral Field Theory," *Speculations in Science and Technology*, **19**, 165-173 (1996).

2. Ginzburg, V.B., "Structure of Atoms and Fields," *Speculations in Science and Technology*, **20**, 51-64 (1997).

3. Ginzburg, V.B., "Double Helical and Double Toroidal Spiral Fields," *Speculations in Science and Technology*, **22** (1998).

4. Ginzburg, V.B., "Nuclear Implosion," *Journal of New Energy*, Vol. 3, No. 4, 1999.

5. Ginzburg, V.B., "Dynamic Aether," *Journal of New Energy*, Vol. 6, No. 1, 2001.

6. Ginzburg, V.B., "Electric Nature of Strong Interactions," *Journal of New Energy*, Vol. 7, No. 1, 2003.

7. Ginzburg, V.B., "Unified Spiral Field Theory – A Quiet Revolution in Physics," *VIA-Vision in Action*, Vol. 2, No. 1 & 2, 2004.

8. Ginzburg, V.B., "The Relativistic Torus and Helix as the Prime Elements of Nature," *Proceedings of the Natural Philosophy Alliance*, Vol. 1, No. 1, Spring 2004.

9. Ginzburg, V.B., "The Unification of Forces," *Proceedings of the Natural Philosophy Alliance*, Vol. 4, No. 1, 2007.

10. Ginzburg, V.B., "The Origin of the Universe, Part 1: Toryces," *Proceedings of the Natural Philosophy Alliance*, The 17th Annual Conference of the NPA 23-26 June 2010 at California State University Long Beach, Vol. 7, 2010.

11. Ginzburg, V.B., "Basic Concept of 3-Dimensional Spiral String Theory (3D-SST)," *Proceedings of the Natural Philosophy Alliance*, The 18th Annual Conference of the NPA, 6-9 July 2011 at the University of Maryland, College Park, USA, Vol. 8, 2011.

12. Ginzburg, V.B., "A Novel Method of Modeling of Fundamental Properties of Materials," Contributed papers from *MS&T15 Materials Science & Technology*, Greater Columbus Convention Center, Columbus, Ohio, USA, October 4-8, 2015.

13. Ginzburg, V.B., "A Novel Method of Modeling of Fundamental Properties of Materials," To be presented at *MS&T17 Materials Science & Technology*, Lawrence L. Convention Center, Pittsburgh, Pennsylvania, USA, October 8-12, 2017.